FIFTH EDITION

THE SKILLED HELPER

Gerard Egan

Gerard Egan, Ph.D., is Professor of Psychology and Organizational Studies at Loyola University of Chicago. He has written over a dozen books, including *The Skilled Helper; Interpersonal Living; People in Systems* (with Michael Cowan); *Moving into Adulthood* (with Michael Cowan); *Change Agent Skills in Helping and Human Service Settings; Change Agent Skills A: Designing and Assessing Excellence; Change Agent Skills B: Managing Innovation and Change;* and *Adding Value: A Systematic Guide to Business-Based Management and Leadership.* He currently writes and teaches in the areas of communication, counseling, business and organization effectiveness, management development, leadership, the management of innovation and change, and organization politics and culture (the "shadow side" of the organization). He also conducts workshops in these areas both in the United States and abroad and is a consultant to a variety of companies and institutions worldwide.

FIFTH EDITION

THE SKILLED HELPER

A Problem-Management
Approach to Helping

Gerard Egan

Loyola University of Chicago

BROOKS/COLE PUBLISHING COMPANY

Pacific Grove, California

I(T)P ™ The trademark ITP is used under license.

 A CLAIREMONT BOOK

Brooks/Cole Publishing Company
A Division of Wadsworth, Inc.

Printed in the United States of America
10 9 8 7 6 5 4 3

Library of Congress Cataloging-in-Publication Data

Egan, Gerard.
  The skilled helper : a problem-management approach to helping /
Gerard Egan. — 5th ed.
    p.   cm.
  Includes bibliographical references and index.
  ISBN 0-534-21294-8
  1. Counseling.   2. Helping behavior.   I. Title.
BF637.C6E39   1994
158'.3 — dc20   93-4291
                                                                  CIP

Sponsoring Editor: *Claire Verduin*
Marketing Representative: *Thomas L. Braden*
Editorial Associate: *Gay C. Bond*
Production Coordinator: *Fiorella Ljunggren*
Production: *Greg Hubit Bookworks*
Manuscript Editor: *Janet Hunter*
Interior Design: *John Edeen*
Cover Design: *Katherine Minerva*
Art Coordinator: *Greg Hubit*
Interior Illustrator: *Kristi Mendola*
Typesetting: *TypeLink, Inc.*
Cover Printing: *Phoenix Color Corporation*
Printing and Binding: *R. R. Donnelley & Sons,*
        Crawfordsville Manufacturing Division

CONTENTS

CHAPTER   3

BUILDING THE HELPING RELATIONSHIP: VALUES IN ACTION   46

CHAPTER   4

A BIAS TOWARD ACTION   66

PART   TWO

BASIC COMMUNICATION SKILLS FOR HELPING   87

CHAPTER   5

COMMUNICATION SKILLS I: ATTENDING AND LISTENING   89

C H A P T E R   6

COMMUNICATION SKILLS II:
BASIC EMPATHY AND PROBING   105

PART THREE

STAGE I OF THE HELPING MODEL AND ADVANCED COMMUNICATION SKILLS   131

STAGE I: HELPING CLIENTS IDENTIFY AND CLARIFY PROBLEM SITUATIONS   133

CHAPTER 7

STEP I-A: HELPING CLIENTS TELL THEIR STORIES   135

CHAPTER 8

STEP I-B: HELPING CLIENTS CHALLENGE THEMSELVES   157

C H A P T E R   9

COMMUNICATIONS SKILLS III: SKILLS AND GUIDELINES FOR EFFECTIVE CHALLENGING   177

C H A P T E R   1 0

STEP I-C: LEVERAGE—HELPING CLIENTS WORK ON THE RIGHT THINGS   199

P A R T   F O U R

HELPING CLIENTS DEVELOP PROGRAMS FOR CONSTRUCTIVE CHANGE   217

C H A P T E R   1 1

PERSPECTIVES AND SKILLS FOR CONSTRUCTING A BETTER FUTURE   219

STAGE III: GETTING THERE — HELPING CLIENTS IMPLEMENT THEIR GOALS   275

CHAPTER 16

STEP III-C: MAKING PLANS—HELPING CLIENTS DEVELOP ACTION PROGRAMS THAT WORK   301

CHAPTER 17

MAKING THINGS WORK—HELPING CLIENTS GET WHAT THEY WANT AND NEED   311

PREFACE

In psychology there has always been tension between practitioners, on the one hand, and both theoreticians and researchers (scientists), on the other. Practitioners face the immediate, and often desperate, needs of their clients on a daily basis; they want to use whatever will work and use it now. Researchers very often focus on such small parts of the helping process that applications to interactions with clients are not immediately evident. Furthermore, there is a tendency among some researchers to argue that their findings are not yet ready to be translated into practice. Finally, some scientists criticize practitioners for using methods that have no scientific basis. Schneider (1990) has claimed that this scenario has led to a lose-lose-lose scenario.

> With rare exceptions, the science and the profession of psychology have failed to give meaning to each other and . . . neither group has attended adequately to the promotion of human welfare. Psychology's empowerments address the private interests of our scientists and our practitioners but they do not address sufficiently the public interest . . . the empowerment of people. . . . We are entrusted with the stuff of what makes people act, think, and feel, and we should be able to provide them with the wherewithal to achieve mastery over their lives and to be effective in society; this would empower us [psychologists] as well. (p. 524)

Scientists lose, practitioners lose; even worse, the clients we serve lose.

Clinical psychologists were supposed to avoid these pitfalls by being trained in the so-called scientist/practitioner model. This model has not worked very well, but we still cling to it. Therefore, there is desperate need in psychology, and especially in the helping professions, for a third role, which I have called *translator*. Translators stay in touch with the best in theory and research and with the needs of practitioners in their service to clients. Their role is to translate the best of theory and research into models, methods, and skills that will benefit practitioners and clients alike. I am sure there are a number of translators out there, even though they don't identify themselves as such. If our academic institutions and internship sites are to produce translators rather than just scientists or just practitioners, then all key players need to buy into the necessity of such translation. There needs to be something of the translator in the scientist and in the practitioner.

This lack of translators or of the translator dimension of scientists and practitioners also leads to the underutilization of some of our robust paradigms. Anderson (1993) has noted that research on human problem solving and human learning have, in the main, been carried out separately. According to him, this is a mistake. Claiming that problem solving is the structure that organizes human thought, he maintains that "research on human problem solving would have been more profitable had it attempted to incorporate ideas from learning theory. Even more so, research on learning would have borne more fruit had Thorndike not cast out problem solving" (p. 35). This lack of synergy has limited the contributions of these two paradigms to the helping professions. In my view, neither problem solving nor learning has been applied as well as possible to counseling and psychotherapy, even though they are, not unexpectedly, two of the most highly researched areas in psychology.

Since the publication of the fourth edition, *The Skilled Helper* has been translated into both Chinese and Japanese. While this proves nothing in itself, it reinforces my belief that problem solving and learning, as core human processes, by necessity underlie *every* approach to helping. If problem solving is not the lead process in any given helping model, then it is the underlying process. As with other models of helping, problem-solving and opportunity-development approaches need to take on the particular cast of the cultures in which they are used. In presenting *The Skilled Helper* model in such places as Africa, Fiji, and Japan, I have always started with the disclaimer: "I will present the model that I use, but you must tell me its relevance to your culture." All have seen the problem-management approach useful — and each has pointed out modifications that are needed in order to adapt it to the local culture. For instance, problem management looks somewhat different in highly authoritarian versus relatively democratic cultures. But even within democratic cultures, there can be significant differences. After I explained what I meant by the term *advanced empathy* to a British audience some years ago, one gentleman raised his hand and said: "We have a term for what you have just outlined and illustrated. We would call advanced empathy all those things that are better left unsaid."

While the search for commonalities among the different approaches to helping continues (Norcross & Goldfried, 1992), problem solving and learning as fundamental integrating mechanisms are given lip service but remain underrepresented in the debate. Furthermore, anthologies of helping models usually ignore problem-solving approaches. This may be because problem solving is seen, however instinctively, as basic to all approaches. This is not to say that any approach to helping, including problem-management approaches, is the elixir of life. A relative of a friend of mine spent years in different kinds of counseling and therapy, including a problem-management approach, to no avail. A couple of years ago, he began taking a mood-altering drug. He has been fine ever since.

The fifth edition of *The Skilled Helper* continues to promote a problem-management and opportunity-development approach to helping and sees counseling and psychotherapy as helper-facilitated processes of client learning. It will probably be a relief to many to know that the basic structure and framework of the helping model itself remain basically intact in this edition. In the past some instructors have winced because of the substantive changes from edition to edition. For that I do not apologize. We should be just as committed to continual improvement of our products and services as the best companies in business and industry currently are. In the fifth edition I have scrutinized every page, paragraph, and sentence in order to simplify the language, rework some of the methods, update the citations, provide new examples, and shorten the book. Stages II and III of the helping model have been recast in terms of a program for constructive change. The language used in these stages is now much more client-focused than process-focused. I have also introduced the notion of the shadow side of the helping process; this is explained at the end of Chapter 1 and illustrated throughout the book. Understanding what is going on in the shadows helps practitioners move from smart to wise.

For offering their comments and suggestions for this edition of *The Skilled Helper*, I would like to thank the following reviewers: John K. Bowers of Northwest Missouri State University, Dale Blumen of the University of Rhode Island, Perilou Goddard of Northern Kentucky University, Thomas Hozman of Shippensburg University, Tracey T. Manning of The College of Notre Dame of Maryland, Stephen D. Miller of Barry University, Dorothy Neufeld of Loma Linda University, Paul H. Raymer of San Diego State University, and Rolfe E. White of the University of Wisconsin at Green Bay. Renewed thanks, as well, to all those who have contributed to previous editions of this book.

Gerard Egan

PART ONE

LAYING THE GROUNDWORK

The centerpiece of this book is the helping model itself; but first there is some groundwork to be laid. This includes outlining the nature and goals of helping (Chapter 1), outlining the helping process (Chapter 2), determining the values that are to drive helping (Chapter 3), and defining the bias toward action that helpers and clients need to develop (Chapter 4).

INTRODUCTION

FORMAL AND INFORMAL HELPERS

Throughout history there has been a deeply embedded conviction that, under the proper conditions, some people are capable of helping others come to grips with problems in living. Today this conviction is institutionalized in a variety of formal helping professions. Counselors, psychiatrists, psychologists, social workers, and members of the clergy are expected to help people manage the distressing problems of life. This book is addressed directly to these professionals and indirectly to a second set of professionals who, although not helpers in the formal sense, often deal with people in times of crisis and distress. Included here are organizational consultants, dentists, doctors, lawyers, nurses, probation officers, teachers, managers, supervisors, and police officers, among others. Although these people are specialists in their own professions, there is still some expectation that they will help their clients or staff manage a variety of problem situations. For instance, teachers teach English, history, and science to students who are growing physically, intellectually, socially, and emotionally and struggling with normative developmental tasks and crises. Teachers are, therefore, in a position to help their students, in direct and indirect ways, explore, understand, and deal with the problems of growing up. Managers and supervisors help workers cope with problems related to work performance, career development, interpersonal relationships in the workplace, and a variety of personal problems that affect their ability to do their jobs.

To all of these professional helpers can be added any and all who try to help relatives, friends, acquaintances, strangers (on buses and planes), and even themselves come to grips with problems in living. Indeed, only a small fraction of the help provided on any given day comes from helping professionals. Peer counselors and helpers, whether or not given these titles, abound in many school, work, church, and social settings. Friends often help one another through troubled times. Finally, all of us must help ourselves cope with the problems and crises of life. Parents must manage their own marital problems while helping their children grow and develop. In short, the world is filled with informal helpers. Since helping and problem solving are such common human experiences, training in both solving one's own problems and helping others solve theirs should be as common as training in reading, writing, and math. But this is not the case.

WHAT HELPING IS ALL ABOUT

Many clients become clients because, either in their own eyes or in the eyes of others, they are involved in problem situations they are not handling well. These clients need ways of dealing with, solving, or transcending their problem situations. In other words, they need to manage their problems in living more effectively.

Problem situations arise in our interactions with ourselves, with others, and with the organizations, institutions, and communities of life.

Clients — whether they are hounded by self-doubt, tortured by unreasonable fears, grappling with cancer, addicted to alcohol or drugs, involved in failing marriages, fired from jobs because they do not have the skills needed in the new economy, suffering from a catastrophic loss, jailed because of child abuse, wallowing in a midlife crisis, lonely and out of community with no family or friends, battered by their spouses, or victimized by racism — all face problem situations that move them to seek help or move others to send them for help.

Some clients come for help, not because they are dogged by problems like those just listed, but because they are not as effective as they would like to be. They have resources they are not using or opportunities they are not developing. People who feel locked in dead-end jobs or bland marriages, who are frustrated because they lack challenging goals, who feel guilty because they are failing to live up to their own values and ideals, who want to do something more constructive with their lives, or who are disappointed with their uneventful interpersonal lives — such clients come to helpers, not to manage their problems, but to live more fully. Many clients, of course, seek helpers because they are not managing problem situations well and because they want to develop their potential and lead fuller lives.

In the end, most clients are not managing their lives as well as they might because they are not grappling effectively with problem situations *and* because they are not taking advantage of unused opportunities. Skilled helpers help them deal with both.

THE GOALS OF HELPING

The goals of helping must be based on the needs of clients. There are two basic goals: one relates to clients' managing their lives more effectively and the other relates to clients' general ability to manage problems and develop opportunities.

Goal One: Problem Management and Opportunity Development

The first goal of helping can be stated in terms of helpers' effectiveness in helping clients move toward more effective problem management.

Helpers are effective to the degree that their clients, through client-helper interactions, are in better positions to manage their problem situations and/or develop the unused resources and opportunities of their lives more effectively.

Notice that I stop short of saying that clients actually end up managing both problems and opportunities better. Since helping is a two-way collaborative process, clients, too, have a primary goal.

Clients are successful to the degree that they capitalize on what they learn from the helping sessions, using these learnings to manage problem situations more effectively and develop opportunities more fully.

In the end, clients can choose to live more effectively or not. Of course, many clients, because of their interactions with helpers, are not only in better positions to manage the ups and downs of their lives more effectively, but actually do so. Counseling is a peculiar kind of service. It is a collaborative process between helper and client. The problem-management and opportunity-development model described in this book is not something that helpers do to clients; it is a process that helpers and clients work through together. In many ways, helpers stimulate clients to provide services to themselves. It is the client who achieves the goals of helping, through the facilitation of the helper. It follows that, although counselors help clients achieve outcomes, they do not control outcomes. When all is said and done, clients have a greater responsibility for both the production and the quality of outcomes.

The Focus on Results

Counseling is an "-ing" word: it includes a series of activities in which helpers and clients engage. These activities, however, have value only to the degree that they lead to valued outcomes in clients' lives. Ultimately, statements such as "We had a good session," whether spoken by the helper or client, must translate into more effective living on the part of the client. If a helper and client engage in the counseling process effectively, something valued will be in place that was not in place before the helping sessions: unreasonable fears will disappear or diminish to manageable levels, self-confidence will replace self-doubt, addictions will be conquered, an operation will be faced with a degree of equanimity, a better job will be found, a woman and man will breathe new life into their marriage, a battered wife will find the courage to leave her husband, and a person embittered by institutional racism will regain self-respect. Helping is about *constructive change*. Skilled counselors help clients develop programs for constructive change.

Consider the case of a battered woman, Andrea N., outlined by Driscoll (1984, p. 64). The mistreatment had caused her to feel that she was worthless even as she developed a secret superiority to those who mistreated her. These attitudes contributed in turn to her continuing passivity and had to be challenged if she was to become assertive about her own rights. Through the helping interactions, she developed a sense of worth and self-confidence. This was the first outcome of the helping process. As she gained confidence, she became more assertive; she realized that she had the right to take stands and chose to challenge those who took advantage of her. She stopped merely resenting them and did something about it. This was a second outcome: a pattern of assertiveness, however tentative in the beginning, took the place of a pattern of passivity. When her assertive stands were successful, her rights became established, her social relationships improved, and her confidence in herself increased, thus further altering the original self-defeating pattern. This was a third set of outcomes. As she saw herself becoming more and more an agent rather than a patient in her everyday life, she found it easier to put aside her resentment and the self-limiting satisfactions of the passive-victim role and to continue asserting herself. This constituted a fourth set of outcomes. The activities in which she engaged, either within the helping sessions or in her day-to-day life, were valuable because they led to these valued outcomes.

Even devastating problem situations can often be handled more effectively. Consider the following example.

> Fred L. was thunderstruck when the doctor told him that he had terminal cancer. He was only 52; death couldn't be imminent. He felt confused, bitter, angry, and depressed. After a period of angry confrontations with doctors and members of his family, in his despair he finally talked to a clergyman who had, on a few occasions, gently and nonintrusively offered his support. They decided to have some sessions together, but the clergyman also referred him to a counselor who worked in a hospice for the dying. With the help of the clergyman and the counselor, Fred gradually learned how to manage the ultimate problem situation of his life. He came to grips with his religious convictions, put his affairs in order, began to learn how to say goodbye to his family and the world he loved so intensely, and, with the help of the hospice in which the counselor worked, set about the process of managing physical decline. There were some outbursts of anger and some brief periods of depression and despair, but generally he managed the process of dying much better than he would have done without the help of his family and his counselors.

This case demonstrates in a dramatic way that the goal of helping is not to solve everything. Fred did die. Some problem situations are simply more unmanageable than others. Helping clients review their problems and the options they have for dealing with them is, as we shall see, a central part of the helping process.

Often opportunity development is as important or more important than problem management. The following case deals with a woman who came for help because she had a problem with lack of commitment and guilt; instead, opportunity development came to center stage.

> After ten years as a helper in several mental health centers, Carol was experiencing burnout. In the opening interview with a counselor, she berated herself for not being dedicated enough. Asked when she felt best about herself, she said that it was on those relatively infrequent occasions when she was asked to help another mental health center to get started or reorganize itself. The counseling sessions helped her explore her potential as a consultant to human-service organizations and make a career adjustment. She enrolled in courses in the Center for Organization Development at a local university. Carol stayed in the helping field, but with a new focus and a new set of skills.

In this case the counselor helped the client manage her problems (burnout, guilt) more effectively through the development of an opportunity that made enormous sense at this stage of her life and of her career.

Goal Two: Helping Clients Become More Effective at Managing Their Lives

The second goal of helping is also client-centered. It deals with clients' ability to continue to manage their lives more effectively after the period of formal helping is over.

Helpers are effective to the degree that clients, through the helping process, learn how to help themselves more effectively.

Just as doctors want their patients to learn how to prevent illness through exercise, good nutritional habits, and the avoidance of toxic activities, just as dentists want their patients to engage in effective prevention activities, so skilled helpers want to see their clients managing *this* problem situation more effectively and to emerge from the helping process capable of more effectively managing subsequent problems in living. Helping at its best empowers clients to become more effective self-helpers.

Counselors can do no better service for their clients than help them become better problem solvers; that is, better problem managers and opportunity developers. Although this book is about a model and methods counselors can use to help clients, more fundamentally, it is about a problem-solving model and methods that clients can use to help themselves. When people are presented with the basic steps of a problem-solving process — such as the stages and steps presented in Chapter 2 — they tend to say "Oh yes, I know that." Recognition of the logic of problem solving, however, is a far cry from using it.

> In ordinary affairs we usually muddle about, doing what is habitual and customary, being slightly puzzled when it sometimes fails to give the intended outcome, but not stopping to worry much about the failures because there are still too many other things still to do. Then circumstances conspire against us and we find ourselves caught failing where we must succeed — where we cannot withdraw from the field, or lower our self-imposed standards, or ask for help, or throw a tantrum. Then we may begin to suspect that we face a problem. . . . An ordinary person almost never approaches a problem systematically and exhaustively unless . . . specifically educated to do so. (Miller, Galanter, & Pribram, 1960, pp. 171, 174)

Clients often are poor problem solvers — or whatever problem-solving ability they have tends to disappear in times of crisis. The function of the helper is to get clients to apply problem solving to their current problem situations and, at the same time, help them adopt more effective approaches to future problems in living.

Helping as a Learning Process

Many people see helping as an education process; I prefer to see it as a *learning* process. Effective counseling helps clients get on a learning track. The helping sessions themselves and the time between sessions involve learning, unlearning, and relearning. Howell (1982) gave us a good description of learning when he said that "learning is incorporated into living to the extent that viable options are increased" (p. 14). In the helping process *learning takes place when options that add value to life are opened up, seized, and acted on*. If the collaboration between helpers and clients is successful, clients learn in very practical ways. They have more degrees of freedom in their lives as they open up options and take advantage of them. This is precisely what counseling helped Andrea N. to do. She unlearned, learned, and relearned and acted on her learnings. Consider another case.

> Tom, a farmer in a midwestern state, lost his farm during a downturn in the economy. At first he was devastated. He did not see himself as a victim of economic forces and government policies beyond his control. He took it personally. In counseling sessions provided by the county, he first worked through his anger at "them" and his guilt about himself. Then he began to learn about himself. He had always believed that he was "meant for the earth," that farming was in his bones. As he was helped to take a closer look at trends in society, he discovered that what he really wanted was not necessarily farming but to be on his own. He got a part-time job, took advantage of some retraining opportunities in computers, and ultimately set up a business helping farmers use computers in their work. He started being what he had always wanted to be without knowing it — an entrepreneur.

Every helping session is an opportunity to help clients identify and act on self-enhancing options. A poet, beset with personal problems, once described himself as "cabined, cribbed, and confined" — a man without options. The helper and the client constitute a learning community. At its best, helping enables clients to learn to throw off chains, to stretch, to open doors, and to go out and find what they want.

Conventional wisdom suggests that experience is the best teacher, but this is not entirely true. Experience *can be* an effective teacher *if* the learner has a model, theory, or framework to use as a tool to extract the learnings that experience has to offer.

DOES HELPING HELP?

Before embarking upon a career in counseling or psychotherapy, you should know that a debate about the usefulness of the helping professions has been going on for at least 35 years (*see* Imber, 1992). Can and do helpers achieve the goals outlined earlier? Cowen (1982) expressed the problem with literary flair.

Once upon a time, mental health lived by a simple two-part myth. Part 1: People with psychological troubles bring them to mental health professionals for help. Part 2: One way or another, often based on verbal dialogue, professionals solve these problems and the people live happily ever after.

And sometimes the cookie does indeed crumble according to the myth. But events of the past several decades suggest that the "marriage-in-heaven" script is not nature's only, or even most frequent, way. In real life the idyllic myth breaks down at several key points.

Let's talk first . . . about Part 2. Heresy though it may have been 20 years ago, it is now permissible to say that not all problems brought to mental health professionals are happily adjudicated. How much of the shortfall is due to the imprecision of our professional "magic," or even to the lack of skill of our magicians, and how much to the selectively refractory nature of the problems that professionals see remains unclear. Much clearer is a sense of mounting dissatisfaction with the reach and effectiveness of past traditional ways, a dissatisfaction that has powered active new explorations toward a more promising tomorrow in mental health. (p. 385)

The literature on the efficacy of helping, whether it is called counseling or psychotherapy, offers us a continuum. At the left end of the continuum is the position "Helping is never helpful." At the right is the position "Helping is always helpful." Neither statement is helpful. There is too much evidence that helping works to side with the nay-sayers and too much evidence that helping often falls short of the mark to side completely with the yea-sayers. After reviewing the evidence, most practitioners would probably say that helping *can* be helpful. In a review of human change process, Mahoney (1991) asks three questions: Can humans change? Can humans help humans change? Are some forms of helping better than others? He gives a qualified "yes" to all three questions. Let's consider two possible reasons why the evidence is not more unambiguously supportive of the efficacy of helping.

The Collaborative Nature of Helping

One reason it is difficult to give a clear-cut answer to the question "Does helping help?" is the collaborative nature of the enterprise. Helping is not like fixing a car or removing a tumor. Helpers do not cure their patients. Helping is at minimum a two-person *team* effort in which helpers need to do their part and clients, theirs. If either party refuses to play the game or to play it well, then the enterprise can fail. Helpers need to give their best to the enterprise. But giving their best means, to a great extent, helping clients tap their own resources: not just getting clients to play the game, as it were, but helping them want to play the game.

Providing counseling or psychotherapy is not the same as giving the client a pill, though sometimes researchers talk as if it were. Outcomes depend on the competence and motivation of the helper, on the competence and motivation of the client, on the quality of their interactions, and often on a host of environmental factors over which neither helper nor client has control. Determining what makes helping effective, then, is much messier than determining whether or not a pill works.

The Competence and Commitment of the Helper

Not all helpers are competent, not all are committed. And even the competent and committed have their lapses. As noted by Luborsky and his associates (1986):

- There are considerable differences between therapists in their average success rates.
- There is considerable variability in outcome within the caseload of individual therapists.
- Variations in success rates typically have more to do with the therapist than with the type of treatment.

Although helping can and often does work, there is plenty of evidence that ineffective helping also abounds. Helping is a powerful process that is all too easy to mismanage. It is no secret that because of inept helpers some clients get worse from treatment. Helping is not neutral; it is "for better or for worse." Ellis (1984) claimed that inept helpers are either ineffective or inefficient. The inefficient may ultimately help their clients, but they use "methods that are often distinctly inept and that consequently lead these clients to achieve weak and unlasting results, frequently at the expense of enormous amounts of wasted time and money" (p. 24). Since studies on the efficacy of counseling and psychotherapy do not usually make a distinction between high-level and low-level helpers, and since the research on deterioration effects in therapy suggests that there are a large number of low-level or inadequate helpers, the negative results found in many studies are predictable.

In the hands of skilled and socially intelligent helpers, helping can do a great deal of good. Kagan (1973) suggested that the basic issue confronting the helping professions is not validity—that is, whether helping helps or not—but reliability:

> Not, can counseling and psychotherapy work, but does it work consistently? Not, can we educate people who are able to help others, but can we develop methods which will increase the likelihood that most of our graduates will become as effective mental health workers as only a rare few do? (p. 44)

To improve the reliability of helping, more effective training programs for helpers are needed.

In the United States, a great deal of the payment for helping services comes from third parties such as government agencies and insurance companies (Winslow, 1993). Because of soaring costs, these institutions are becomingly increasingly results oriented. The progress of treatment and the effectiveness of helpers are being tracked. In one tracking system, clients are given a "mental health index" score after an initial assessment. The score indicates such things as symptom severity and ability to function in family, social, and work settings. Studies showed that clients who don't show significant improvement over four sessions are not likely to benefit from further treatment. Third parties are putting pressure on the helping industry to demonstrate that helping helps and that it is cost-effective.

You are encouraged to acquaint yourself with the ongoing debate concerning the efficacy of helping (Freedheim, 1992; Hill & Corbett, 1993; Horvath & Symonds, 1991; Kendall, Kipnis, & Otto-Salaj, 1992; Smith, Glass, & Miller, 1980; Steketee & Chambless, 1992; Weiss, Weiss, & Donenberg, 1992). Study of this debate is not meant to discourage you, but to help you (1) appreciate the complexity of the helping process, (2) acquaint yourself with the issues involved in evaluating the outcomes of helping, (3) appreciate that, poorly done, helping can actually harm others, (4) make you reasonably cautious as a helper, and (5) motivate you to become a high-level helper, learning and using practical models, methods, skills, and guidelines for helping.

A PROBLEM-MANAGEMENT/ OPPORTUNITY-DEVELOPMENT APPROACH TO HELPING

In the face of all this diversity, helpers, especially beginning helpers, need a practical, working model of helping that enables them to learn:

- what to do to help people facing unmanaged problems in living,
- how to help clients develop unused resources and opportunities,
- what specific stages and steps make up the helping process,
- what techniques aid the process,
- what communication skills are needed to interact with clients,
- how these skills and techniques can be acquired,
- what clients need to do to collaborate in the helping process and to manage their problems and develop their opportunities more effectively, and
- how to evaluate their own and their clients' efforts.

A flexible, humanistic, broadly based problem-management and opportunity-development model or framework meets all of these requirements. It is precisely such a model that is presented in the pages of this book—a model that is straightforward without ignoring the complexities of clients' lives or of the helping process itself. Such a model is a tool for learning. Given away to clients, it is a tool clients can use to learn from their experiences both outside and inside helping sessions.

Common sense suggests that problem-solving models, techniques, and skills are important for all of us, since all of us must grapple daily with problems in living of greater or lesser severity. Ask anybody whether problem-management skills are important for day-to-day living, and the answer inevitably is "certainly." Talk about the importance of problem solving is everywhere. Yet if you review the curricula of our primary, secondary, and tertiary schools, you will find that talk outstrips practice. There are those who say that formal courses in problem-solving skills are not found in our schools because such skills are picked up through experience. To a certain extent, that's true. However, if problem-management skills are so important, why does society leave their acquisition to chance? A problem-solving mentality should become second nature to us. The world may be the laboratory for problem solving, but the skills needed to optimize learning in this lab should be taught; they are too important to be left to chance. A problem-management model in counseling and therapy has the advantage of the vast amount of research that has been done on the problem-solving process itself. The model, techniques, and skills outlined in this book tap that research base.

An Open-Systems Model

The problem-management model is an open-systems model, not a closed school. It takes a stand on how counselors may help their clients, yet it is open to being corroborated, complemented, and challenged by any other approach, model, or school of helping. In this respect it fits the description of efficient therapy advanced by Ellis (1984) when he said that efficient therapy "remains flexible, curious, empirically-oriented, critical of poor theories and results, and devoted to effective change. It is not one-sided or dogmatic. It is ready to give up the most time-honored and revered methods if new evidence contradicts them. It constantly grows and develops. . ." (p. 33). The needs of clients — not the egos of model builders — must remain central to the helping process.

Universality — Problem Management as an Underlying Process

Library shelves are filled with books on counseling and psychotherapy — books dealing with theory, research, and practice. In the 1980s it was estimated that there were between 250–400 different approaches to helping (*see* Karasu, 1986). And the number continues to grow. If you were to leaf through books that are compilations of the different schools or approaches to counseling and psychotherapy, you would soon discover a bewildering number of schools, systems, methods, and techniques, all of which are proposed with equal seriousness and all of which claim to lead to success. Ultimately, however, all models converge around the principal client-centered goal of managing problems in living more effectively and developing opportunities more fully. Therefore, *the problem-management and opportunity-development process outlined in this book underlies or is embedded in all approaches to helping.* Problem management and opportunity development spans not only all schools and models of helping but also all cultures. Even though the problem-management and opportunity-development process takes different forms in different cultures, its basic structure is recognized around the world.

Systematic Eclecticism: Choosing the Best Practice

The eclectic helper borrows good ideas from a variety of approaches and molds them into his or her own approach to counseling (Jensen, Bergin, & Greaves, 1990; Lazarus, Beutler, & Norcross, 1992). Most helpers borrow in this way to a greater or lesser extent; indeed, research has shown that most counselors and psychotherapists like to consider themselves eclectic (Norcross & Prochaska, 1988). However, an effective eclecticism must be more than a random borrowing of ideas and techniques from here and there. Helpers need a conceptual framework that enables them to borrow ideas, methods, and techniques systematically from all theories, schools, and approaches and integrate them into their own theory and practice of helping (Mahalik, 1990). The helping model in this book is a tool for borrowing and integrating ideas that work, no matter what their origin. Clients are not interested in theories; they are interested in outcomes.

A comprehensive problem-management model can be used to make sense of the vast literature in counseling and psychotherapy in at least three ways — mining, organizing, and evaluating:

1. *Mining.* Helpers can use the problem-management model to mine any given school or approach, digging out whatever is useful without having to accept everything else that is offered. The stages and steps of the model serve as tools for identifying methods and techniques that will serve the needs of clients.

2. *Organizing.* Since the problem-management model is organized by stages and steps, it can be used to organize the methods and techniques that have been mined from the rich literature on helping. For instance, a number of contemporary therapies have elaborated excellent techniques for helping clients identify blind spots and develop new perspectives on the problem situations they face. As you shall see as we discuss Step I-B, the "new perspectives" step of the problem-management model, these techniques can be organized.

3. *Evaluating.* Since the problem-management model is pragmatic and focuses on outcomes of helping, it can be used to evaluate the vast number of helping techniques that are constantly being devised. The model enables helpers to ask in what way a technique or method contributes to the bottom line; that is, how it contributes to outcomes that serve the needs of clients.

There is a movement afoot to pool all the good ideas found in the many different forms of counseling and psychotherapy in order to identify a set of converging themes: the principles, approaches, and methodologies that constitute the essence of helping (Andrews, 1991; Goldfried & Castonguay, 1992; Kottler, 1991; Norcross, Alford, & DeMichele, 1992; Norcross & Goldfried, 1992; Okun, 1990). This approach moves one step beyond eclecticism, even systematic eclecticism, because it focuses on the essence of helping. However, some helpers fear that one helping system would be developed and then imposed on everyone, with the result that the richness that comes with diversity would be lost. The integration movement is still in its infancy, and, as you can imagine, is plagued with professional politics. It is an important movement, however, and one worth watching. The problem-management and opportunity-development model embedded in all forms of helping also serves as a model for convergence.

WHAT THIS BOOK IS — AND WHAT IT IS NOT

"Beware the person of one book," we are told. For some people the central message of a book becomes a cause; religious books such as the Bible or the Koran are examples. Although causes empower people, we should beware when a person believes that all he or she needs to know is in one book. Such a person remains closed to new ideas and growth. Certainly all truth about helping cannot be found in one book. Therefore, *The Skilled Helper* is not and cannot be "all that you've ever wanted to know about helping."

It is a pragmatic model of helping.   Since this book cannot do everything, it is important to state what it is meant to do. Its purpose is to provide helpers — whether novices or those with experience — a practical framework or model of helping and some of the methods and skills that make the model work. It is designed to enable helpers to engage in activities that will help their clients manage their lives more effectively. In short, this book is part of the helping curriculum — an extremely important part — but it is not the whole.

It is not the total curriculum.   Clients are the customers of helpers and have every right to expect the best of service from them. Beyond a helping model and the skills that make it work, what kind of training enables helpers to "deliver the goods" to our clients? A practical curriculum is one that enables helpers to understand and work with their clients in the service of problem management and opportunity development. The curriculum includes both working knowledge and skills. "Working knowledge" is the translation of theory and research into the kind of applied understandings that enable helpers to work with clients. "Skills" refer to the actual ability to deliver services.

A fuller curriculum for training professional helpers includes:

- a working knowledge of applied developmental psychology (Fassinger & Schlossberg, 1992; Gibson & Brown, 1992);
- the application of the principles of human behavior to the helping process (Watson & Tharp, 1993);
- the principles of cognitive psychology as applied to helping (Freeman & Dattilio, 1992; McMullin, 1986; Persons, 1989);
- applied personality theory;
- abnormal psychology;

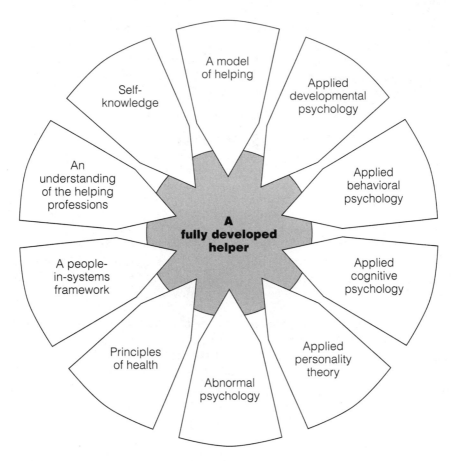

FIGURE 1-1
A Sample Curriculum for Helpers

- the principles of social psychology (Heppner & Frazier, 1992);
- an understanding of health and wellness and of clients as psychosomatic beings (Altmaier & Johnson, 1992; Myers, Emmerling, & Leafgren, 1992);
- an understanding of clients as "people-in-systems" (Egan & Cowan, 1979; Egan, 1984, L. A. Gilbert, 1992); that is, an understanding of clients in the social settings of their lives, an understanding of the diversity of clients (Pedersen, 1991b, 1991c);
- an understanding of the needs and problems of special populations with which one works, such as the elderly or substance abusers; and
- an understanding of the dynamics of the helping professions themselves (Fretz & Simon, 1992).

While much of this curriculum is touched on indirectly, especially through the many examples offered, this book is in no way a substitute for such a curriculum. Figure 1–1 outlines a sample curriculum for helpers.

MANAGING THE SHADOW SIDE OF HELPING

Throughout this book the helping relationship and process are described and illustrated in very positive terms. They are described as they should be, not as they always are. The shadow side of helping can be defined as: *All those things that adversely affect the helping relationship and process, its outcomes, and its impact in substantive ways but that are not identified and explored by helper or client.* The ability to understand how helper, client, the relationship, and the helping process itself can go wrong is the first step toward managing the shadow side. For instance, helpers' motives are not always as pure as they are portrayed in this book. Helpers, even though incompetent, pass themselves off as professionals. Some are not very committed even though they are in a profession in which success demands a high degree of commitment. Clients often play games with themselves, their helpers, and the helping process. Helpers seduce their clients; clients seduce their helpers, not necessarily sexually. Hidden agendas are pursued by helpers and clients. The helping relationship itself can end up as a conspiracy to do nothing. Clients, aided and abetted by their helpers, work on the wrong issues. Helping methods that are not likely to benefit *this* client are used. Though decisions are made throughout the helping process, decision making itself remains murky. Helping is continued even though it is going nowhere. The list goes on.

But this is not an exercise in cynicism. Clearly, not all of these things happen all the time. To assume so would be cynical. The shadow side is not usually an exercise in ill will. Few helpers set out to seduce their clients. Few clients set out to seduce their helpers. Helpers don't realize that they are incompetent. Clients don't realize that they are playing games. Most clients most of the time have what Schein (1990) called a "constructive intent" and most helpers do their best to put their own concerns aside to help their clients as best they can. But this does not mean that any given helper is always giving his or her best. Clients and helpers have their temptations. Clients, as we shall see, have their blind spots; helpers have their own.

All human endeavors have their shadow side. Companies and institutions are plagued with internal politics and are often guided by covert or vaguely understood beliefs, values, and norms that do not serve their best interests. If helpers don't know what's in the shadows, they are naive. If they believe that shadow-side realities win out more often than not, they are cynical. Helpers should be neither naive nor cynical. Rather they should pursue a course of upbeat realism.

It is certainly naive to assume that everyone with a problem wants to be helped. It is also naive, it would seem, to assume that everyone can be helped. Sometimes, as Kierulff (1988) demonstrated in the following case, the shadow side takes the form of the personality structure of the client.

A young man (whom I will call John) attempted to rob a bar. He and his older partner carried loaded shotguns, but the bartender was armed and quick on the draw, and John was shot. The . . . bullet severed John's spine; his legs collapsed under him, and he was left paraplegic. When he was taken to the hospital, he was assigned to me for psychotherapy.

John and I talked for hours, day after day, as he lay prone in his hospital bed. This 20-year-old armed robber fascinated me. He lied to me straightfaced. He portrayed himself as a victim. He changed his story whenever he thought he could gain an advantage. . . . He displayed no loyalty, no honor, no compassion. He trusted no one, and he displayed not even a crumb of trustworthiness. He used everyone, including me.

I had not dealt with any sociopaths before interacting with John. . . . I believed that Rogerian unconditional positive regard would eventually work its wonders and soften John's tough shell, allowing me to connect with him on a level of mutual empathy, caring, and respect.

"You can't change these people," my supervisor admonished. . . . I hoped that my supervisor was wrong. I kept on trying, but when the internship was over and I said goodbye to John, I could tell from the look in his eyes that in spite of the extra concern I had shown him, in spite of my warmth and care and effort, I was just another in the long list of people he had coldly manipulated and discarded. (p. 436)

In the ensuing article, Kierulff mused on the meaning of and the relationship between free will and determinism and what kind of treatment could get through to "guiltless, manipulative people." He certainly learned that helpers who allow themselves to be conned are doing no good. He also learned that the ultimate success of counseling is never in the hands of the helper.

The tone of this book is unabashedly upbeat. However, helpers-to-be must not ignore the less palatable dimensions of the helping professions, including the less palatable dimensions of themselves. Therefore, throughout this book some of the common shadow-side realities that plague client, helper, and the profession itself are noted at the service of managing them. This book is by no means a treatise on the shadow side; it should, however, get helpers to begin to think about the shadow side of the profession. Skilled helpers are idealistic without being naive. They know the difference between realism and cynicism — and they opt for the former.

CHAPTER 2

OVERVIEW OF THE HELPING MODEL

A NATURAL PROBLEM-MANAGEMENT PROCESS

There is a natural decision-making process that helping processes can assist, modify, and accelerate but not replace. Understanding and respecting this natural process and its variations is part of the wisdom of helping. Yankelovich (1992), a public-opinion expert, outlined a process that applies to public opinion formation and decision making, especially as related to tough issues. Applied to problem management and opportunity development, the natural steps look something like this.

1.   Clients become *aware* of an issue or set of issues. For instance, not just this or that marital problem but vague dissatisfaction with the relationship itself.

2.   A *sense of urgency* develops, especially as the problem situation becomes more annoying or painful. This and the first step constitute the consciousness-raising stage of the process. To continue our example, subsequent annoyances in the marriage are now seen in light of overall dissatisfaction.

3.   Clients begin to *look for remedies*. However perfunctorily, different strategies for managing the problem situation are explored. Clients in difficult marriages begin thinking about complaining openly, separating, getting a divorce, instituting subtle acts of revenge, having an affair, going to a marriage counselor, seeing a minister, unilaterally withdrawing from the marriage in one way or another, and so forth.

4.   The *costs of pursuing different solutions* begin to emerge. "If I confront my partner openly, I'll have to go through the agony of confrontation, denial, argument, counteraccusations, and who knows what else." Or, "What would I do if I were to go out on my own?" Or, "What would happen to the kids?" The possibilities here are endless and clients often retreat because there is no cost-free or painless way of dealing with the problem situation.

5.   Since the problem situation is now seen for what it is, it is impossible to retreat completely and so *a more serious weighing of choices* takes place. For instance, the costs of confronting the situation are weighed against the costs of merely withdrawing. Often, a kind of dialogue goes on in the client's mind between steps four and five.

6.   An *intellectual decision* is made to accept some choice and pursue a certain course of action. "I'm going to bring all of this up to my spouse and suggest we see a marriage counselor." Or, "I'm going to get on with my life, find other things to do, and let the marriage go where it will." A merely intellectual decision, however, is often not enough to drive action.

7.   The *heart joins the head* in the decision. One spouse might finally say, "I've had enough of this!" That is, the decision is permeated by values and emotion. This fuller decision is more likely to drive action. A spouse might say, "It is unfair to both of us to go on like this; it just isn't right," and this drives the decision to seek help, even if this means going alone.

Note that this natural problem-managing process can be derailed at almost any point along the way. For instance, the costs of managing the problem (the fourth step) might seem too high and the problem situation is put on the back burner.

The problem-management process outlined in this chapter and developed in the rest of this book can be used to provide focus for this natural process, to add other steps that are helpful but not as natural, to challenge backsliding, and to speed up the process. But it *complements* this natural process rather than taking its place. In other words, the *technology* of helping must respect human nature.

Moreover, since the natural process described evolves differently in different individuals, individual differences must also be respected. For instance, some people remain in the denial phase longer than others. Finally, since helpers may enter this process at any step along the way, they need to determine just where the client is. If the client is in the looking-for-remedies phase, helpers should not immediately assume that the client needs to return to earlier steps.

Helpers who unwittingly assume that their theories, models, and methods replace these natural processes are in trouble. Ignoring this natural process or assuming that it applies to every client in every situation the same way is part of the shadow side of helping. Respecting and complementing this natural process, on the other hand, is the basis for client-centered helping.

CLIENT-CENTERED HELPING

Clients who are having difficulty managing problem situations and/or developing unused opportunities constitute the starting point of the helping process. In this book, client-centered helping means that the needs of the client are the starting point, not the models and methods of the helper. It is an *active* approach, however, and often includes, as we shall see later, challenging clients — or inviting them to challenge themselves — to explore problems and opportunities, set goals, and actively pursue them. The ultimate focus is, of course, client-enhancing outcomes.

Problem situations.   Clients come for help because they have crises, troubles, doubts, difficulties, frustrations, or concerns. These are often generically called "problems," but they are not problems in a mathematical sense, because often emotions run high and often there are no clear-cut solutions. It is probably better to say that clients come, not with problems, but with problem situations; that is, with complex and messy problems in living that they are not handling well. They come for help because they want to or because third parties, concerned about or bothered by their behavior, send them. For instance, Duane sees a clergyman because he is despondent after discovering his wife is having an affair with a colleague; Jason is sent to see a counselor because of behavioral problems in the classroom; a battered wife willingly accepts help from a mental health center; and her husband is forced into counseling because the court mandates it.

Missed opportunities and unused potential.   Clients' missed opportunities and unused potential constitute a second starting point for helping. In this case, it is a question not of what is going wrong, but of what could be better. It has often been suggested that most of us use only a fraction of our potential. Most of us are capable of dealing much more creatively with ourselves, with our relationships with others, with our work life, and, generally, with the ways in which we involve ourselves with the social settings of our lives. Counselors can help clients empower themselves by helping them identify and develop unused or underused opportunities and potential. Klaus, for example, seeks help because he has heard over and over that he is a talented and engaging person but seems to do little with his talents. The fact that he is moving into his 40s, the "mid-life crisis" time, acts as a stimulus.

Mixed cases.   In most cases, counselors interact with clients who are grappling with difficulties and failing to take advantage of opportunities. Even when clients talk about problems, effective helpers listen for missed opportunities and unused potential. Kelly, a woman filled with self-doubt because of her sheltered upbringing and the obstacles presented by a male-dominated culture, has missed opportunities and failed to develop many of her own resources. She may well be a victim of society, but she may also be a victim of her own unexploited potential.

THE SKILLED-HELPER MODEL

A helping model is like a map that helps you know what to do in your interactions with clients. At any given moment, it also helps you to orient yourself, to understand where you are with the client and what kind of intervention would be most useful. There are three "stages" together with a bias toward action. I put "stages" in quotation marks because, as we shall see at the end of the overview, this term has sequential overtones that are somewhat misleading. In practice the three stages overlap and interact with one another as clients struggle through the natural process of constructive change just outlined. At any rate, these stages outline what *clients* need to do in order to manage a problem situation or develop an opportunity. This book outlines what helpers can do to *assist* clients as they move through these stages. Figure 2-1 shows an overview of the main facets of the skilled-helper model.

Stage I: Reviewing the current scenario.   *The Principle: Help clients identify, explore, and clarify their problem situations and unused opportunities.* Clients can neither manage problem situations nor develop opportunities unless they

The skilled-helper model

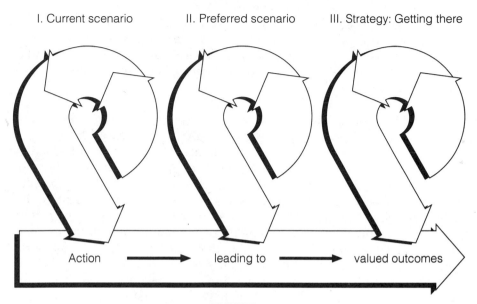

I. Current scenario   II. Preferred scenario   III. Strategy: Getting there

Action ➡️ leading to ➡️ valued outcomes

FIGURE 2-1
Overview of the Helping Model

identify and understand them. Initial exploration and clarification of problems and opportunities takes place in Stage I. As indicated in Figure 2-1, Stage I deals with the current scenario—a state of affairs that the client or those who send the client for help find unacceptable. Problem situations are not being managed and opportunities are not being developed. For instance, a counselor helps Carla, single and underemployed, discuss her concerns about her aging mother with failing health and their deteriorating relationship.

Stage II: Developing the preferred scenario.   *The Principle: Help clients identify what they want in terms of goals and objectives that are based on an understanding of problem situations and opportunities.* Once clients understand either their problem situations or opportunities for development more clearly, they may need help in determining what their options are. Stage II deals with the preferred scenario. The preferred scenario provides answers to such questions as: "What do you want? What would things look like if they were better?" This stage deals with outcomes, results, what clients want. Carla determines that things would be better if she and her mother were living separately and if she were to develop a fuller personal life, including a better job and a more active social life.

Stage III: Getting there.   *The Principle: Help clients develop action strategies for accomplishing goals, for getting what they want.* Clients may know what they want to accomplish and where they want to go, but still need help in determining how to get there. This stage deals with activities, ways of achieving results. The counselor helps Carla explore ways of establishing a different set of living arrangements with her mother. He also helps her to identify and focus on the guilt she experiences as she discusses these arrangements. The counselor also helps her find ways of getting a better job and of establishing a wider social life.

Getting things done: A bias toward action. *The Principle: Help clients act on what they learn throughout the helping process; help clients translate strategies into goal-accomplishing action.* These three stages are *cognitive* in nature—talking about and planning for change. Talk, however, must be translated into action. In many helping or problem-management models, action or implementation is tacked on at the end. In this model, all three stages sit on the action arrow, indicating that clients need to act in their own behalf right from the beginning of the counseling process. As we shall see, clients need to act within the helping sessions and in their real day-to-day worlds. Carla finds the sessions more useful if she comes prepared. Each time she arrives with an agenda. She also finds the sessions more useful if she leaves them with a clear idea of things she must do between sessions. In the end, Carla gets her brothers and sisters to help her find a good nursing home located near one of her older brothers and his family. She also enrolls in a nursing program and starts dating.

Each stage of the model has three steps. Be aware though, that these steps are not always sequential. Clients don't take one step after another. Rather they move back and forth among the stages and steps toward the accomplishment of problem-managing goals. Effective helping tends to be fluid and flexible. In practice the problem-management and opportunity-development process is not as clean, clear, and linear as described in these pages. Remember that the model is a framework or map to help you find your way as a helper. It is also an outline of this book.

STAGE I: REVIEWING PROBLEM SITUATIONS AND UNUSED OPPORTUNITIES

In this overview, the highlights from an actual case will be used to illustrate each step of the model. The case has been disguised and simplified. It is not a session-by-session presentation but is presented here to illustrate the stages of the helping model. The client is voluntary, verbal, and, for the most part, cooperative. In actual practice, cases do not flow as easily as this one. The simplification of the case, however, will help you see the entire model in action.

I. Current scenario   II. Preferred scenario   III. Strategy: Getting there

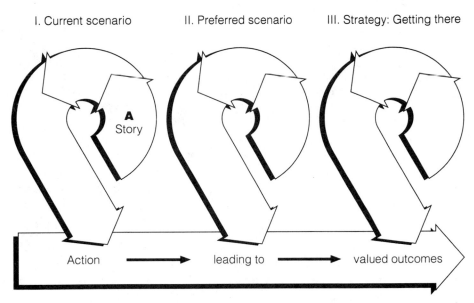

FIGURE 2-2
The Helping Model: Step I-A

Step I-A: Telling and Clarifying the Story

The Principle: Help clients tell their stories as clearly as possible. Helpers cannot be of service if clients fail to develop an understanding of the difficulties and possibilities of their lives. Therefore, as indicated in Figure 2-2, clients need to tell their stories; they need to reveal and discuss their problem situations and their missed opportunities. Some clients are quite verbal; others may be almost mute. Some clients easily reveal everything that is bothering them; others are quite reluctant to do so. Involuntary clients often prefer to talk about the failings of those who sent them.

Helpers need skills that enable them to help clients discuss their problem situations and to provide support for them as they do so. The outcome of this step, then, is a frank discussion of the facts of the case. The story needs to be told, whether it comes out all at once at the beginning of the helping process or only in bits and pieces over its entire course. To do this, of course, helpers need to establish effective relationships with their clients.

Ray, 41, was a member of a research team for a large agricultural firm. The team's research focused on the development of improved planting seed. He came to see me at the suggestion of a mutual friend because he was experiencing a great deal of stress. In our first meeting I helped Ray tell his story, which proved to be somewhat complex. He was bored with his job, his marriage was lifeless, he had poor rapport with his two teenage children, he drank too much, his self-esteem was low, and he had begun to steal small things, not because he needed them but because he got a kick out of taking them and "getting away with it." He told his story in a rather disjointed way, skipping around from one problem area to another. His agitation in telling the story was apparent but did not get in his way. He was embarrassed, and a nonjudgmental attitude and supportive communication style on my part helped him overcome the uneasiness he was experiencing, at least to a degree.

Since Ray was a talented person, I assumed that missed opportunities for growth were also a part of the overall picture, even though he did not talk about these directly. Ray was a voluntary client and quite verbal; much of the story came tumbling out in a spontaneous way. Involuntary clients, clients with poor verbal skills, or clients who are hostile, confused, ashamed, or extremely anxious present a much greater challenge.

Step I-B: Identifying and Challenging Blind Spots

The Principle: Help clients discover and deal with the kinds of blind spots that keep them from seeing problems and opportunities clearly and moving ahead. One of the most important things counselors can do is help clients identify blind spots and develop new, more useful perspectives on both problem situations and unused opportunities. As shown in Figure 2-3, this is the second step of Stage I. Most clients need to move beyond their initial subjective understanding of their problem situations. Many people fail to cope with problems in living or to exploit opportunities because they do not see them from new perspectives. Comfortable but outmoded frames of reference keep them locked into self-defeating patterns of thinking and behaving. Helping clients get rid of "blinkers" that stand in the way of problem management and opportunity development is one of the major ways in which counselors can empower clients, to help clients unleash and use the power they have. Challenging blind spots is not the same as telling people what they are doing wrong. Rather, it is helping people see themselves, others, and the world around them in more creative ways.

When Ray talked about his family life, he tended to blame his wife, Laura, and his children for his woes. In time I invited him to take a deeper look at these problems, noting that in most families there are few pure-form victims and few pure-form troublemakers. Ray at one point admitted that both he and Laura let the marriage stagnate. At one time he half mused and half asked: "Wouldn't it seem that most marriages end up bland?" This helped refocus him on what he wanted from his marriage. As Ray told his story, he described himself as "underemployed" at one time and as "over the hill" at another. I helped him confront the discrepancy between these two descriptions of himself. As we discussed this issue, Ray came to see that he had always disliked people he saw as overly ambitious. But he also began to see that he had moved to the other end of the spectrum and had become underambitious in self-defeating ways.

I. Current scenario   II. Preferred scenario   III. Strategy: Getting there

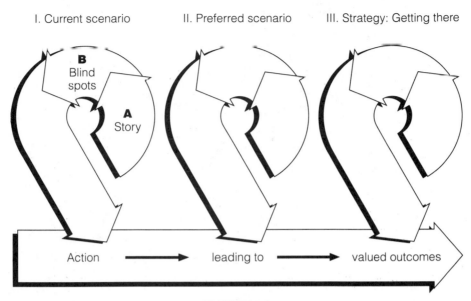

FIGURE 2-3
The Helping Model: Step I-B

Effective helpers realize that support without challenge is often superficial and that challenge without support can be demeaning and self-defeating. It is important to understand the client's frame of reference or point of view even when it is evident that it needs to be challenged or at least broadened. Finally, challenge at the service of new perspectives, though presented here as a step, is best woven into the fabric of the entire helping process. Helping clients challenge themselves with respect to goals, action strategies, and action itself is part of the process.

Step I-C: Searching for Leverage

The Principle: Help clients identify and work on problems, issues, concerns, or opportunities that will make a difference in their lives. In Figure 2-4 the term *leverage* is used as the generic name of this third step. First, if clients, in telling their stories, reveal a number of problem situations at the same time, as Ray does, or if the problem situation is complex, then it is important to help the client determine which one or which part to work on first. In other words, counselors help clients establish priorities and search for some kind of leverage in dealing with complex problem situations. Second, since helping is an expensive proposition, both financially and psychologically, some kind of screening is called for. Is the chosen problem, at least as stated, worth the time and effort that is about to be spent in managing it?

It was clear from the beginning that Ray had some substantive issues to work on. Since he tended to wander from topic to topic, I suggested that he could

I. Current scenario   II. Preferred scenario   III. Strategy: Getting there

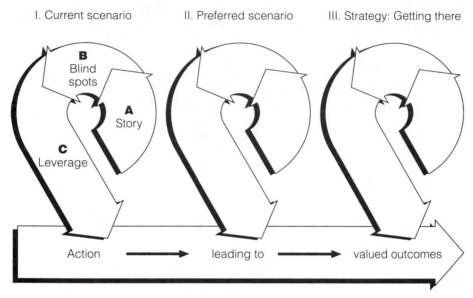

FIGURE 2-4
The Helping Model: Step I-C

pick one issue and review it in more depth. Ray did this. It proved that most of the issues he mentioned in the first session were interrelated anyway. After some discussion, he decided that he would like to tackle the job problem. With the help of a few probes and prompts, Ray discussed his dissatisfaction with his current job and the impact this had at home. I had a suspicion that he had chosen an easier issue, but it was his choice and the other issues were still on the table.

Some of the principal work-related issues were clarified: Ray saw himself in a dead-end run-of-the-mill research job. He felt, on reflection, that he had always been underemployed, that his talents had never been significantly appreciated or challenged. I spent time understanding Ray's point of view and, on occasion, inviting him to take a deeper look at himself. For instance, it is not just *they* who have not appreciated and challenged him; he has not done this for himself. At any rate, his dislike for his job had given him an attitude that was contaminating his interpersonal relationships at work and adding to his stress. He saw that his negative feelings about his job were also carrying over into his home life. Ultimately, I pointed out that he talked about his problems mainly in terms of what was happening to him and not in terms of what he did or failed to do. This was a tough pill to swallow because he hated victims.

The problem Ray chose to work on touched on most of the others. Skipping all over the place and getting nothing done would not have been a viable option, but there were different ways of helping Ray in his search for leverage. Clients need not lock themselves into dealing with the first problem or opportunity they choose to work on. Remember, helping at its best is not disorganized, but is fluid and flexible. There is no formula. The principle remains: help clients deal with issues that make a difference.

Client Action: The Heart of the Helping Process

The Principle: Help clients act both within and outside the counseling sessions. Helping is ultimately about constructive change. Talking about problems and opportunities, discussing goals, and figuring out strategies for accomplishing goals is just so much talk without action. There is nothing magic about change; it is hard work. If clients do not act in their own behalf, nothing happens. Two kinds of client action are important: actions within the helping process itself and actions in the client's day-to-day world. Ideally, by their actions, clients come to "own" the helping process instead of being the objects of it. The fact that Ray had come voluntarily was a good sign. He was acting in his own behalf. He was in pain and wanted to do something about it. His willingness to explore problem situations and respond fairly nondefensively to a helper's challenges on the part of a helper were indications that he was taking ownership of the process.

But taking an active role within sessions is only the first step. One reason that Ray was in trouble was that he had consistently failed to act on these issues. It was easier to blame others. He had let life pass him by in a variety of ways and now was angry with himself and his world. The stages and steps of the helping process are, at best, triggers and channels for client action. Client actions between sessions — not random actions, obviously, but actions leading to client-enhancing outcomes — are the best indicators of how well the sessions are going. It makes no sense to say "We really had a good session today!" if the session bears no fruit in real life. Perhaps the best predictor of ultimate success in helping is what clients do between sessions.

> Although I learned this only later, after our first session Ray sat down one evening, wrote the word *drinking* at the top of a piece of paper, and jotted down, randomly, whatever thoughts came to his mind. The paper began to fill up with such things as money spent on liquor, times spent drinking, attempts to keep others from knowing, occasions when drinking had kept him from work or led to family fights, and so forth. He found all of this painful and eventually ripped the page up and threw it into a wastebasket, saying to himself that drinking was probably no more a problem for him than for anyone else "in his situation." Obviously, even though Ray had not chosen drinking as one of the first issues to be discussed, it was still on his mind.

This was not a dramatic act with a dramatic outcome, but at least Ray did *something* after the first session. He took one somewhat ambiguous and abortive step in exploring his drinking. In the beginning some client actions may be merely symbolic — Ray's list comes close to that. But little actions related to substantive issues help create the climate of action that moves the helping process forward.

Stage II: Developing the Preferred Scenario

People often take a one-stage approach to managing their problems. That is, when something goes wrong, they act, or rather, they react. For instance, Troy, a college student, found himself lonely, frustrated, and depressed during his first week away from home. When another student, noting his misery, not only expressed some concern but also intimated a sexual attraction, Troy was confronted with his own sexual ambiguity. He "solved" this complex problem situation by running away and getting a job in a hotel in another city. Obviously his action, however understandable because of his panic, solved nothing.

Others take a two-stage approach to managing their problems. They engage in some kind of analysis of the problem and then come up with solutions based on this analysis. For instance, Nina, whose mother had been kidnapped from the parking lot of a mall, taken to a secluded place, raped, and murdered, was supposed to go away to college. She felt guilty leaving her father and younger sister so soon after the tragedy. After talking with her parish priest, she decided to "do the right thing" and stay home. As time passed, she realized that her solution didn't solve anything.

In the skilled-helper model, there are three stages. There is something between analyzing a problem or unused opportunity and doing something about it. This middle stage deals with desired or preferred outcomes. For instance, Nina, her father, and her sister could have been helped to identify what they wanted, what needed to be in place in order to manage the problem situation. In its most complete sense, the word *solution* refers to outcomes or accomplishments, to what will be *in place* once a set of actions are taken. Stage II deals with what clients want, with their preferred outcomes, not with what they need to do to get what they want.

Step II-A: Developing Preferred-Scenario Possibilities

The Principle: Help clients develop a range of possibilities for a better future. If a client's current state of affairs is problematic and unacceptable, then he or she needs to be helped to conceptualize a new state of affairs — that is, alternatives, more acceptable possibilities. As indicated in Figure 2-5, this is the first step in Stage II. In Step II-A, helpers, against the background of the problem identification and clarification of Stage I, help clients develop answers to the questions "What do really I want? What would my life look like if it looked better?" Both research and everyday experience show that very few of us, when confronted by a problem or an unexploited opportunity, ask ourselves such questions as "What would this problem situation look like if managed? What would be in place that is not now in place? What

I. Current scenario   II. Preferred scenario   III. Strategy: Getting there

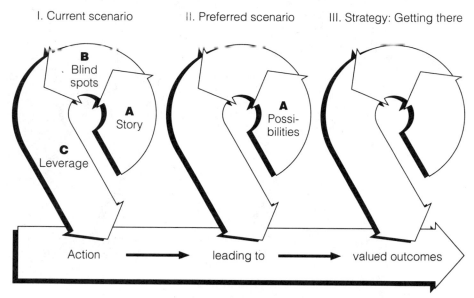

FIGURE 2-5
The Helping Model: Step II-A

would this opportunity look like if it were developed? What would exist that does not exist at this moment?" These are powerful questions because they help clients use their imaginations and provide direction by helping them focus on a better future rather than a frustrating present.

A new scenario is not a wild-eyed, idealistic state of affairs, but rather a picture of a better future.

> Ray finally realized that doing something about his marriage was very important. Laura agreed to come to the sessions. We did not spend a lot of time reviewing the past and I did not let them play the "who's to blame" game. With a bit of guidance, they quickly moved to describing what a better future might look like. Some possibilities included greater mutual respect, less indifference, better communication, more openness, more effectively managed conflicts, a more equitable distribution of household tasks, some kind of career for her, becoming a team in relationship to their children, surrendering past animosities, a renewed sex life, and less game playing. There was some understandable skepticism on both their parts as they reviewed these possibilities, but they managed to stay clear of cynicism.

The failure to imagine possibilities different from the present contributes a great deal to stagnation in the helping process. In this stage, "What do you want?" is a powerful question. Later clients can be helped to become realistic about what is possible, but here the freeing of the imagination is critical.

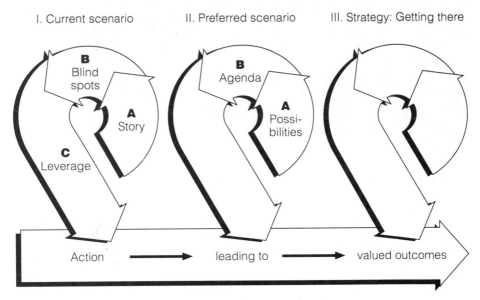

I. Current scenario   II. Preferred scenario   III. Strategy: Getting there

B Blind spots

A Story

C Leverage

B Agenda

A Possibilities

Action   →   leading to   →   valued outcomes

FIGURE 2-6
The Helping Model: Step II-B

Step II-B: Translating Possibilities into Viable Goals

The Principle: Help clients choose realistic possibilities and turn them into viable goals. The possibilities developed in Step II-A constitute the building blocks for the desired outcomes of the helping process. "What do you want?" gives way to "What do you really want?" Figure 2-6 adds this step to the helping model. Goals chosen by the client need to be viable; they must be capable of being translated into action.

Since Ray and Laura wanted better communication as part of the renewal of their marriage, they needed to spell out what better communication meant to them. What undesirable patterns of communication were in place? What kinds and patterns of communication were preferred? They both agreed that when they experienced small annoyances in their relationship, they tended not to share them with each other but to swallow them. Well, they *thought* they swallowed them, but in reality they both saved them up and let them eat away inside. Ray at first said that this is "what women do." Later he admitted he did the same thing. The new pattern of communication was to include discussing annoyances as they arose, and doing so without being petty or without playing put-down games. As Ray said, "If we were to manage little problems better, we could avoid the bigger ones." Laura agreed: "Our marriage hasn't exploded; it's imploded."

In a couple of one-on-one sessions Ray explored the job satisfaction issue. While it was true that he was underemployed, he now saw that this was mainly his problem, not his company's. He reviewed various job possibilities,

I. Current scenario   II. Preferred scenario   III. Strategy: Getting there

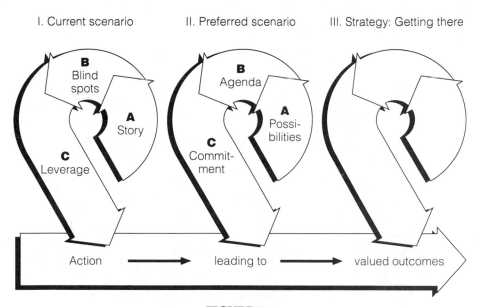

FIGURE 2-7
The Helping Model: Step II-C

and even though he was a bit impatient in discussing the pros and cons of each job, he did so. The last thing he needed was to make an improper career change. In the end he chose to stay with his company but to get a job in an entirely different department—one dealing with biotechnology, the cutting edge of the company's business. His goal was not just a new job but a new approach to work. The new job would include supervising other technicians. At first Ray had not wanted to be "responsible for anyone else." Now he was actually intrigued about the challenge of not just doing good research but "playing a role in making good research happen."

Choosing from among possibilities can be difficult for clients because it often involves painful self-scrutiny and choices. Sometimes the choices themselves may be severely limited. In patients with terminal illnesses, choices between almost certain death and medical interventions such as agonizing chemotherapy sessions or a series of operations are not easy to make.

Step II-C: Commitment to a Program of Constructive Change

The Principle: Help clients identify the kinds of incentives that will help them to pursue their chosen goals. What are clients willing to pay for what they want? Choosing goals is one thing. Pursuing them is another. The more appealing the goals, the better; however, appealing or not, commitment is essential. Figure 2-7 adds this commitment step to the helping model. Since one of the problems everyone faces in the pursuit of goals is competition from other agendas, help from counselors at this step is useful, and particularly so when the choices are hard. What are the incentives for such a choice?

> Even after Ray had committed himself to getting a new, more challenging job inside the firm, his search never seemed to get off the ground. Instead, he seemed to develop some renewed interest in the job he was in. His problem drinking, which had almost disappeared, popped up again. His relationship with Laura, which had improved appreciably, got ragged at times. I said to him, "We both know you're dragging your feet. What's going on?" As much as he wanted the new position, he was afraid to go for it. He feared he would be turned down. He winced when he thought of his new responsibilities. The thought of being responsible for others in a supervisory position had lost much of its luster. Our talk helped. The very fact that he had kept his qualms to himself compounded the problem. Talking about the up side of the new job helped him get back on track. His misgivings did not completely disappear, but his procrastination did.

His incentives included a strong desire to get rid of the pain he was experiencing and a more positive desire to develop some of the talents that had remained dormant over the past years. Another incentive was his conviction that getting a more fulfilling job would help enormously in doing his part to put his marriage back on course. A final incentive was his need to regain his self-esteem.

If Stage II is done well, clients will have a clear idea of what they want and where they would like to go — even though they might not know how to get there. Ray and Laura still had to struggle with the how of accomplishing their package of marriage-reinvention goals. This brings us to Stage III of the helping process.

STAGE III: DETERMINING HOW TO GET THERE

Preferred-scenario goals constitute *what* clients want, where clients want to go. Stage III deals with *how* to get there.

Step III-A: Brainstorming Strategies for Action

The Principle: Help clients brainstorm a range of strategies for accomplishing their goals. In this step, added in Figure 2-8, clients are helped to ask themselves (with apologies to the poet), "How can I get where I want to go? Let me count the ways!" Some clients, when they decide what they want, implement the first strategy that comes to mind. While the bias toward action may be laudable, the strategy may be ineffective, inefficient, imprudent, or a combination of all three. At this stage of the problem-managing process, as

I. Current scenario   II. Preferred scenario   III. Strategy: Getting there

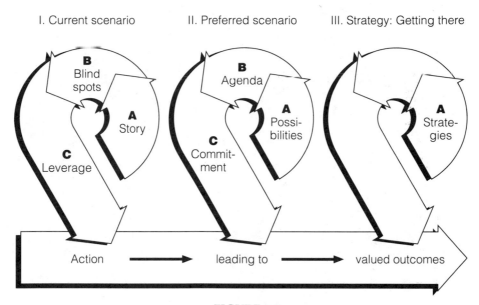

FIGURE 2-8
The Helping Model: Step III-A

many strategies as possible (within time and other constraints) should be uncovered. Even seemingly outlandish strategies can provide clues for realistic action programs. Let us return to Ray.

> In one session Ray and Laura are asked to "wing" answers to the question, "What are the thousand things you need to do to be better communicators?" At first they think it's silly, but once they get into the swing of things, they have a great deal of fun coming up with possible strategies. But even some of the outlandish possibilities end up being helpful. One wilder possibility was to do a TV show together showing others how an effective marriage works by showing them their own. In the end, after getting involved with a movement called Marriage Encounter, they end up as group facilitators. Not exactly a TV show, but the TV idea opened doors.

Doing what was necessary to change jobs was somewhat easier. After our "Why are you dragging your feet?" session, Ray found out what he needed to do to apply for the job and he did it. He thought there might be stumbling blocks. But his skills, however dormant, were well known, and his enthusiasm carried the day. He even said, "The job wasn't the problem. I was." It made him muse on why he tended to sell himself short, mainly to himself.

Step III-B: Choosing the Best Strategies

The Principle: Help clients choose a set of strategies that best fit their environment and resources. Once a number of different options for action have been

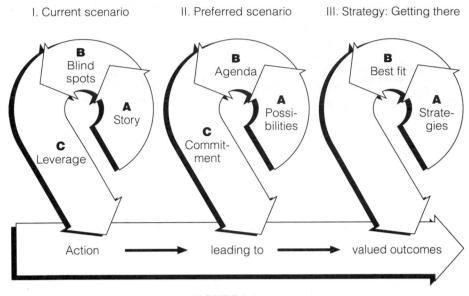

FIGURE 2-9
The Helping Model: Step III-B

identified, then client and helper collaboratively review them and try to choose the best; that is, the single strategy or combination of strategies that best fits the client's needs, preferences, and resources and that is least likely to be blocked by factors in the client's environment. In other words, strategies for accomplishing goals need, like goals themselves, to be evaluated in terms of their realism. The options chosen must also be in keeping with the values of the client. Figure 2-9 adds this step to the helping model.

> Ray and Laura reviewed some of the strategies they had come up with to improve their communication. One was to take a course in communication skills together at a local university. However, they eliminated this possibility for several reasons, including cost, time constraints, and the possibility of having to air some of their marital issues with younger students. On the other hand, involvement in the Marriage Encounter movement made a great deal of sense. The expense was reasonable, they could fit meetings in around their schedules, and the church affiliation of the group dovetailed with another one of their strategies, some kind of renewal of their religious roots.

Realism is important because there is no sense choosing strategies that can lead only to failure. Helpers need to be careful at this point, however. Sometimes clients choose strategies that make them stretch. Such a choice is fine, provided that the stretch is within reason. In the end, the Marriage Encounter movement provided both Ray and Laura with a great deal of stretch.

I. Current scenario   II. Preferred scenario   III. Strategy: Getting there

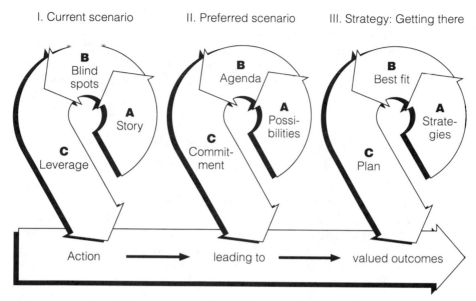

FIGURE 2-10
The Helping Model: Step III-C

Step III-C: Turning Strategies into a Plan

The Principle: Help clients formulate a plan, a step-by-step procedure for accomplishing each goal of the preferred scenario. Clients are more likely to act if they know what they are going to do first, what second, what third, and so forth. The strategies that are chosen need to be translated into a step-by-step plan. Realistic time frames also add value. Plans help clients impose self-enhancing discipline on themselves. Figure 2-10 adds this step to the helping model.

Let's see how this step applies to Ray.

Once Ray had switched to his new job and he and Laura had begun to get involved with the Marriage Encounter groups, we had what we called a planning session. They brought a calendar and I helped them sketch out what the coming year might look like. We all noted that the very fact that they were now planning things together was a critical strategy for the communication-improvement goal. In their enthusiasm they tended to schedule more events — Marriage Encounter meetings, get-away weekends, involvement in parish activities, and the like — than seemed reasonable to me. We discussed the difference between a "stretch" program for constructive change and one that might overtax their resources. In the end, they adopted a quite realistic approach. They started with a wish list of things they wanted to do. Then at the beginning of each month they translated items on the wish list into fixed commitments. This way they were able to factor in such things as the changing demands of Ray's job.

Self-responsibility is a key value in Stage III. Ray and Laura had the responsibility for pulling together their own plans. I added value in a variety of ways—helping them clarify items on their list, challenging them to make the distinction between a stretch program and a pie-in-the-sky one, and getting them to discuss frankly the extent to which *both* of them owned the events they were scheduling.

ACTION REVISITED: PREPARING AND SUPPORTING CLIENTS

The function of planning is to institute and give direction to problem-managing and opportunity-developing action. The actions clients must take to implement constructive-change programs constitute the transition phase of counseling. This is illustrated in Figure 2-11, where the action arrow is now called transition. There are a number of things counselors can do to help clients prepare themselves for and engage in the implementation of action programs. First, counselors can help clients in their immediate preparation for action. This may be called the "forewarned is forearmed" phase. Effective counselors help clients foresee difficulties that might arise during the actual execution of their plans. In this regard there are two extremes. One is pretending that no difficulties will arise. The client optimistically launches into a program and then runs headlong into obstacles and fails. The other is spending too much time anticipating obstacles and figuring out ways of handling them. This can be just another way of delaying the real work of problem management and opportunity development. Second, helpers can also provide support and challenge for clients during the implementation of the action programs themselves. Clients need both to support and challenge themselves and to find support and challenge from others. The helper can also challenge the client to mobilize whatever resources are needed to stick to a program.

> In getting a different job within his company, Ray realized that his full goal was more than just getting the job. He wanted to renew his image within the company and with himself. Just before he moved to the new job, we discussed this fuller picture and identified pitfalls and obstacles. One possible obstacle was his new boss. He needed the job filled but was not convinced that Ray was the perfect candidate. Since Ray knew this, we talked about what needed to be done to establish himself with his new boss. Ray decided that "giving more than 100%" in the initial weeks was the way to go. He and Laura agreed that he would have to spend some long hours learning the ropes. But these hours were seen as investment rather than cost.
>
> As Ray and Laura began implementing their program for reinventing their marriage, they ran into a few glitches. Laura, having more time, wanted to give more time to the Marriage Encounter movement. Ray saw too many competing agendas. Instead of having them work this issue out with me, I sug-

I. Current scenario   II. Preferred scenario   III. Strategy: Getting there

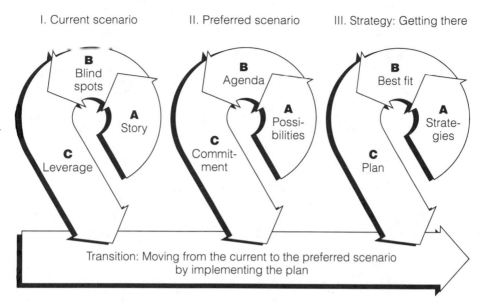

FIGURE 2-11
The Transition from the Current to the Preferred Scenario

gested that they take advantage of the resources at the Marriage Encounter center with some of their peers. Other couples had experienced similar difficulties and could share their hard-earned wisdom. Also, since the Marriage Encounter was now a part of their lives, it put the helping process more into real time. It was time for me to bow out of the process anyway.

Ray found no perfect solution to his personal problems nor did he and Laura reinvent the perfect marriage. At the end Ray was still, on occasion, experiencing problems with drinking. Working separately and together, however, they did both manage their problem situations more effectively and develop their potential more fully. Although I worked with them over a three-year period, I did not spend that much time with them. Some of the sessions with Ray lasted only ten minutes. The sessions with Ray and Laura together lasted, on average, about 30–40 minutes.

ONGOING EVALUATION OF THE HELPING PROCESS

In many helping models, evaluation is presented as the last step in the model. However, if evaluation occurs only at the end, it is too late. Therefore, in Figure 2-12, an "E" is placed in the center of every stage to indicate that the helping process needs to be evaluated throughout. That is, helpers

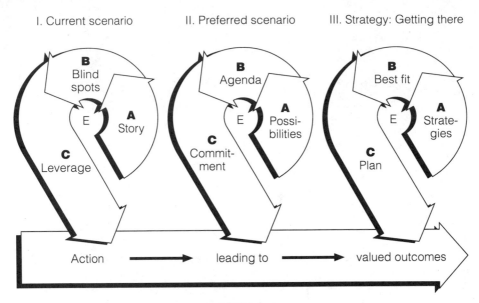

FIGURE 2-12
Evaluation in the Helping Model

and clients need to ask themselves "In what ways are the counselling sessions contributing substantially to problem management and opportunity development?" Questions are provided in this book to help you and your clients engage in ongoing evaluation.

FLEXIBILITY IN THE USE OF THE MODEL

The overview in this chapter has presented the logic of the helping model. The case presents the "literature" of the model; that is, the model in actual use. I realize, however, that this case may be too pat. It does not give the full flavor of the hard work Ray and Laura did or the turmoil they experienced at times. It does not give a full picture of the raggedness of helping. The sole purpose of this example, however, is to provide a picture of the model in action and some idea of how you can use it to give direction to the entire process.

The helping model, while providing guidance, must remain flexible to the needs of clients. One form of rigidity is to drag clients mechanically through the stages and steps of the model. This is not how clients work. For instance, since clients do not always present all their problems at once, it is impossible to work through Stage I completely before moving on to Stages II and III. New problems must be explored and understood whenever they are presented. The model must also respond to the realities of the actual helping situation. Helpers will often find themselves moving back and forth in

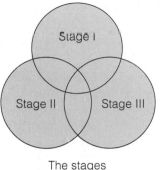

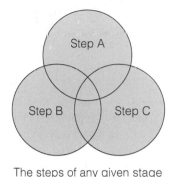

The stages   The steps of any given stage

FIGURE 2-13
Overlap of Stages and Steps

the model. A client might actually try out some action strategy even before the problem situation is adequately defined and some goal has been set. This is not necessarily a bad thing. Action sometimes precedes understanding. If the action is not successful, then the counselor helps the client learn from it and return to the tasks of clarifying the problem situation and setting some realistic goals. One, two, or more steps or even two stages of the process often merge into one another. For instance, clients set goals and develop strategies to achieve them at the same time. New and more substantial concerns arise while goals are being set, and the process moves back to an earlier exploratory stage. In practice, then, both the stages and the steps in the model overlay and mingle with one another, as illustrated in Figure 2-13.

All of this underscores the importance of having a map. The model is the map. Since the stages and steps of the model are often jumbled together, it helps immeasurably to have a map that shows where we have come from, where we are now, and where we need to go.

BECOMING A SKILLED HELPER

As the title of this book suggests, helpers-to-be need more than a helping model; they need a range of skills to make the model work. These include basic and advanced communication skills; the ability to establish working relationships with clients and to help clients challenge themselves; skills in problem clarification, goal setting, development of an action plan, implementation of the plan, and ongoing evaluation. These reflect creative and humane ways of being with clients in their efforts to develop a program of constructive change. The only way to acquire these skills is by learning them experientially, practicing them, and using them until they become second nature.

Basic Steps of the Training Process

A full program for learning the skilled-helper model includes the following basic steps:

1. *A conceptual understanding of the steps of the model.* Reading this book, reviewing examples, and listening to lectures will give you a conceptual or cognitive understanding of these skills.
2. *A behavioral understanding of the helping model.* By watching instructors model these skills, by watching videotapes of effective helpers, and by doing the exercises in the manual that accompanies this book, you will develop a behavioral rather than just a conceptual feeling for these skills.
3. *Initial mastery.* You will begin to master these skills by actually practicing them with your fellow trainees under the supervision of a trainer. In this stage you will at times feel awkward using your newly-learned skills.
4. *Further mastery.* You will strengthen your hold on these skills by using them in practicum or internship experiences under the supervision of skilled and experienced helpers. Awkwardness in the use of helping skills lessens as they begin to become part of you.
5. *Lifelong learning.* The best helpers continue to learn throughout their careers from their own experiences, from the experiences of their colleagues, from reading, from seminars, and — perhaps most importantly — from their interactions with their clients. In this sense, full mastery is a journey rather than a destination.

Case Management

Chapter 1 included a brief discussion about what this book was and what it was not. Now that we have reviewed the helping model itself, an additional comment is in order. This book does not deal directly with case management: it does not include specific ways of dealing with *this* kind of client in *this* kind of situation. Yet, some suggestions related to case management are included and others are made indirectly through the many examples used to illustrate the helping process. The pragmatics of case management are more effectively learned through the training and supervision program outlined earlier.

The helping model just reviewed, however, is one of the most effective tools in case management. The issues clients bring with them can be called *content* problems. The surgery patient talks about her fears of dying, a married couple talk about their seemingly irreconcilable differences, and a sex-abuse victim talks about his disoriented life. Content problems deal with the clients' lives. But clients can also have problems with the helping process itself. These can be called *process* problems. Clients are belligerent because they have been sent, they keep missing sessions, they won't talk about key issues, they are reluctant, they are resistant, there seems to be some incompatibility between client and helper, they are too anxious to collaborate in the helping process — the list goes on.

The helping model itself can be used to manage such process problems. Consider the following case.

- **Current scenario**:   The client is avoiding talking concretely about the principal issues related to her problem situation.
- **Preferred scenario**:   The client is talking more freely and concretely about her principal concerns and this is helping her begin to manage her problems.
- **Strategies for getting there**:   The helper brainstorms strategies for helping the client talk more freely and chooses the strategies that make most sense.
- **Action**:   The helper uses these strategies to manage the process problem.

In this example, the key is devising and using strategies to help the client overcome the process problem. One client might respond to a direct challenge: "The issues you're talking about are very sensitive, and I appreciate how difficult it may be for you to talk about them in any detail. Is there anything that I can do to help you do this more easily?" Another might be helped by a more indirect strategy:

HELPER: Let's change roles for a little while. You are me and I am you. If you were the counselor, what would you be saying to me right now?
CLIENT: Let me think. . . . I'd say that you seem to be stalling around. Not talking about the real issues, you know, and that's getting in the way.

Sometimes in frustration or even panic we can fail to use our own medicine in managing the problems we encounter as helpers. In training counselors, I inevitably run into endless what-if questions ("What if the client does x, y, or z. Then what do I do?") or very real problems trainees have run into in practicum and internship situations ("The client did x, and I didn't know what to do"). When the trainee is challenged to apply the three-stage model to the process problem, almost inevitably she or he comes up with a number of strategies for handling it. There can be hundreds of different process problems. When I throw a problem open to the rest of the class, even more, and often more creative, strategies are identified. It is impossible to come up with a specific strategy for dealing with each problem, just as it is impossible to devise a single strategy for all the hundreds of content problems clients present. The key is to use the problem-solving model to generate appropriate ways of getting to the preferred state of affairs.

Understanding and Dealing with the Shadow Side of the Helping Process

Besides the broad shadow-side themes mentioned in Chapter 1, there are a number of shadow-side pitfalls in the use of any helping model, including the constructive-change model used here. They include, among others, rigidity, overcontrol, virtuosity, and ineptness.

Rigidity.   The model discussed and illustrated in this book provides the basic principles of helping others manage their lives more effectively. It does not provide easy formulas. The model itself is not a linear formula. A beginning helper, or even an experienced but unskilled one, can apply the helping model too rigidly. Helping is for the client; the model exists only to aid the helping process. Flexibility is essential. Helpers can drag clients in a linear way through the model, making sure that clients get help in all stages when, in reality, the clients may need only selective help. Counselors need to focus on the big picture and not the details of the model. Helping should always be client-driven rather than model-driven.

Overcontrol.   Helping models can provide focus and direction, thus helping clients develop self-discipline in the management of their problems in living. On the other hand, since helping models are powerful tools, they can be used to control clients rather than helping them liberate themselves.

Virtuosity.   Some helpers tend to specialize in the skills of a particular stage or step of the model in a formulaic way. Some specialize in problem exploration; others claim to be good at challenging; others want to move immediately to action programs. Helpers who specialize not only run the risk of ignoring client needs but often are not very effective even in their chosen specialties. For example, the counselor whose specialty is challenging clients is often an ineffective challenger. The reason is obvious: challenge must be based solidly on an understanding of clients. Effective counselors have a wide repertoire of skills and use them in a socially intelligent way. These skills enable them to respond spontaneously to a wide variety of client needs. In sum, counseling is for the client; it is not a virtuoso performance by the helper. Clients and helper alike create the formulas needed as they move along. This makes helping a client-centered process.

Ineptness.   Using a model and using it well are not the same thing. Any model in the hands of an inept helper is a dangerous thing. Inept helpers do not necessarily cause harm, but they can make the helping process inefficient. How long does helping take? How long should it take? There is no formula. A general rule, however, is that the helper should move as quickly through the stages and steps needed by the client as the client's abilities and state permit. The client should not be penalized for the helper's lack of skills. Beginning helpers often dally too long in Stage I, not merely because they have a deep respect for the necessity of relationship development and problem clarification, but because they either do not know how to move on or fear doing so. High-level clients may be able to move quickly to action. Their helper, obviously, should be able to move with them.

CHAPTER 3

BUILDING THE HELPING RELATIONSHIP: VALUES IN ACTION

THE HELPING RELATIONSHIP

VALUES IN ACTION — HUMANIZING THE HELPING PROCESS
Pragmatism — Whatever Is Ethical and Works
Competence — Adding Value
Respect — Valuing Diversity and Individuality
Genuineness — Beyond Professionalism and Phoniness
Client Self-Responsibility — Nonpatronizing Empowerment

HELPING AS A SOCIAL-INFLUENCE PROCESS
A Brief History of Social Influence in Helping
Reconciling Self-Responsibility and Social Influence

THE CLIENT-HELPER WORKING CHARTER

THE SHADOW SIDE OF VALUES

THE HELPING RELATIONSHIP

Theoreticians, researchers, and practitioners alike—not to mention clients—all agree that the helping relationship is very important (Gaston, 1990; Grencavage & Norcross, 1990); not all agree why. Some stress the relationship in itself (*see* Bailey, Wood, & Nava, 1992; Kahn, 1990); others stress the work that is done through the relationship (Reandeau & Wampold, 1991), and still others stress the outcomes achieved through the relationship (Horvath & Symonds, 1991). Few studies have focused on the ability of clients to establish working relationships with their helpers. One study showed that clients with antisocial personality disorders (clients whose track records in helping situations would be very poor) may benefit to the degree that they can participate in a working alliance (*see* Gerstley et al., 1989).

The relationship itself.   Patterson (1985) made the relationship itself central to helping: "Counseling or psychotherapy is an interpersonal relationship. Note that I don't say that counseling or psychotherapy involves an interpersonal relationship—it is an interpersonal relationship" (p. 3). If that is the case then taking pains to establish, develop, and foster the relationship is also central. Deffenbacher (1985) outlined what these activities are:

> Good . . . therapists work to build rapport, lessen interpersonal anxiety in the relationship, increase trust, and build an interpersonal climate in which clients can openly discuss and work on their problems. Their clients need to perceive that they have a caring, positive, hopeful collaborator in understanding and making changes in their world. Good listening in an open, nondefensive manner, careful attention and tracking of client concerns, direct, honest feedback, and the like are not only important for assessment, but also create the context in which further exploration and [helping] strategies may be implemented. Clients . . . need to feel cared for, attended to, understood, and genuinely worked with if successful therapy is to continue. (p. 262)

Stressing the relationship itself is especially important when helping is seen as a forum for social-emotional reeducation. In this view, it is through the relationship that clients begin to care for, trust, and challenge themselves. A helper at the termination of the helping process might be able to say of her client,

> Because I trusted him, he trusts himself more; because I cared for him, he is now more capable of caring for himself; because I invited him to challenge himself and because I took the risk of challenging him, he is now better able to challenge himself. Because of the way I related to him, he now relates better both to himself and to others. Because I respected his inner resources, he is now more likely to tap these resources.

The relationship as a means to an end.   Others see the helping relationship as extremely important but primarily as a means to an end. A good relationship enables client and counselor to collaborate in the use of the helping model and the methods that make it work. The relationship is subservient and instrumental to achieving the goals of the helping process. Some say that the importance of the relationship cannot be overstressed. This, I believe, is a mistake. The relationship itself must not obscure the ultimate goal of helping: clients' managing their lives better. This goal won't be achieved if the relationship is poor, but if too much focus is placed on the relationship itself, both client and helper are distracted from the real work to be done.

A client-centered working alliance.   The term *working alliance* is a good one. First, it underscores the idea that helping entails hard work on the part of both client and helper. Second, the term *alliance* denotes a pragmatic partnership, one related to results. The idea of the one perfect kind of helping relationship is a myth. Different clients have different needs, and these needs are best met through different kinds of relationships. One client may work best with a helper who expresses a great deal of warmth; another might work best with a helper who is more objective and businesslike. Some clients come to counseling with a fear of intimacy: they might be put off by helpers who, right from the beginning, communicate a great deal of empathy and warmth. Effective helpers use a mix of skills and techniques tailored to the kind of relationship that is right for each client.

VALUES IN ACTION — HUMANIZING THE HELPING PROCESS

Values have always been important in the helping process (Bergin, 1991; Beutler & Bergan, 1991; Kerr & Erb, 1991; Norcross & Wogan, 1987). Since it has become increasingly clear that helpers' values influence clients' values over the course of the helping process, it is essential to build a value orientation into the process itself (Vachon & Agresti, 1992). However, we must be aware that values cannot be handed to prospective helpers on a platter. The focus of this chapter is broader than simple values; it covers the beliefs, assumptions, values, norms, and standards that serve as the philosophical and moral foundations of helping. What is written here is meant to stimulate your thinking about values. You need to be proactive in your search for the beliefs, values, and norms that will govern your interactions with your clients. This does not mean that you will invent a set of values different from everyone else's. We all learn from the tradition of the profession and the wisdom of others. The set of values you come up with will certainly be influenced by the values of your clients, of the helping profession, and of society, together with whatever transcendent values you see governing interactions among people. Ultimately, however, you must commit yourself to your own package of values. For, as you sit with your clients, only those values that you have made your own will make a difference.

Values are not just ideals. In the pragmatics of helping others, values are a set of criteria for making decisions. That is true of yourself as a helper and of clients as they grapple with problems and opportunities. For instance, a helper might say to himself or herself,

> The arrogant I'm-always-right attitude of this client needs to be challenged. How to challenge her is another story. Since I respect the client, I do not want to challenge her by putting her down. On the other hand, I value openness. Therefore, I can challenge her by describing the impact she has on me and do so without belittling her.

Working values enable the helper to make decisions on how to proceed. Helpers without a set of working values are adrift.

Values that guide transactions between helpers and their clients can be packaged in a variety of ways. The five major sets addressed here are pragmatism, competence, respect, genuineness, and client self-responsibility.

Pragmatism — Whatever Is Ethical and Works

Any kind of helping worth its salt is useful, right from the start. It is also based on sound ethical standards. The chapter focuses more on values than ethics. This is not to say that morality and ethics in helping are not important. On the contrary (American Psychological, 1992; London, 1986; Pope & Vasquez, 1991), ethics and morals are so important that helpers-to-be are urged to review ethical principles and applications both in some formal setting such as a course and on their own. An ethical and humane pragmatism can be expressed in a variety of ways.

Maintain a real-life focus. If clients are to make progress, they must do better in their day-to-day lives. The focus of helping, then, is not narrowly on the helping sessions and client-helper interactions themselves, but on clients' managing their day-to-day lives more effectively. In his early days as a helper, a friend of mine exulted in the solid relationship he was building with a client until the client stopped during the third interview, stared at him, and said, "You're really filled with yourself, aren't you? But, you know, we're not getting anywhere." He had become so lost in relationship building that he forgot about the client's pressing everyday concerns.

Stay flexible. As noted earlier, this book suggests a framework, principles, methods, and skills, but no formulas. The entire process of helping needs to be adapted to the condition and needs of the client. Move with the client. The flexibility principle is: Do whatever is ethical and works. A friend of mine held one of his sessions with a client in a shopping mall. The client needed to get back into community and the quiet one-to-one isolation of the office setting underscored apartness. Flexibility is not an invitation to do stupid things or to turn helping into theater. It is an invitation to tailor the process to the client.

Develop a bias toward action. A pragmatic bias toward acting — rather than merely talking about acting — is a cardinal value. I am active with clients and see no particular value in mere listening and nodding. I engage clients in a dialogue. During that dialogue, I constantly ask myself, "What can I do to raise the probability that this client will act in her own behalf more intelligently and prudently?" I know a man who went into therapy years ago because, among other things, he was indecisive. Over the years he has become engaged several times and each time has broken it off. So much for decisiveness.

Do only what is necessary. Helping is an expensive proposition, both monetarily and psychologically. Even when it is "free," someone is paying for it through tax dollars, insurance premiums, or good-will offerings. Therefore, without rushing nature or your clients, get to the point. Do not assume that you have a client for life. Helping can be lean and mean and still be most human. A colleague of mine experimented, quite successfully, with shortening the counseling "hour." I do not assume that the average helper will get to the point he did with his research subjects when he would begin by saying, "We have five minutes together. Let's see what we can get done." He was very respectful, and it was amazing how much he and his clients could get done in a short time.

Be realistic. Part of pragmatism is realizing the limitations of your profession. First, most people muddle by without professional help. Second, clients can get better in a variety of ways without your help. Third, professional help is not meant for everyone. I often show a popular series of videotaped counseling sessions in which the client receives help from a number of helpers, each with a different orientation. Once the course participants have seen the tapes, I ask, "Does this client need counseling?" They all practically yell, "Yes!" Then I ask, "Well, if this client needs counseling, how many people in the world need counseling?" Just because a person might well benefit from counseling does not mean that he or she needs counseling. Finally, helping that achieves only partial results may, at times, be the best that we can do. Ray and Laura, whose case we considered in Chapter 2, did not end up with a perfect marriage. But, working together, they realized that their relationship was worth saving and improving — and they did create a better marriage.

Competence — Adding Value

Studies show that competent helpers do help; that is, they directly add value to the client's program for constructive change and indirectly add value to the client's life. On the other hand, incompetent helpers usually are not neutral; they actually do harm. Studies also show that helpers with university degrees — supposed professionals — are not necessarily competent. On the other hand, some helpers without degrees are very competent. Informal and paraprofessional helpers, even without degrees, can be professional in the sense that they actually do help others manage their lives more effectively. Competence involves the following norms.

Become good at helping. Become proficient in delivering the helping model described and illustrated here. Being able to deliver the goods to clients goes far beyond merely understanding models and having an intellectual understanding of what needs to be done. Helpers need plenty of supervised experience in real-life helping situations. Counselor training programs should, in my opinion, be competency-based. Certificates and degrees should be awarded only if the helper can deliver the goods. One way of becoming good is practice. You can use the problem-management model outlined in this book every day in your own life. The best learners immerse themselves in it until it becomes second nature.

Continue to learn. I once gave a talk to a professional psychology club in a large city. Since most of the people there were professionals, I did not want to talk down to them. During the question period, however, it soon became apparent that many of them had not been keeping up with the profession. This does not automatically mean that they were incompetent. But look at it this way: How comfortable would you feel going to a doctor who had not read a medical journal for years? The curriculum outlined briefly in Chapter 1 is not just for school, but for life.

Practice what you preach. Demonstrate your competence by modeling the kinds of things you challenge clients to do. If you want them to be open, be open. If you want them to act, do not be afraid to be active in the helping sessions. One client told me that she was not going to go back to her counselor at the university counseling center because "He just sits there and nods. I can get anyone to do that." Effective helpers use the tools of their trade on their own problems.

Be assertive. If you are good at what you do, don't apologize for it. Do it. Many trainees in the helping professions are afraid to assert themselves. Some want to settle for a caricature of nondirective counseling as a helping model, not necessarily because they espouse the theory underlying it but because their interpretation (or misinterpretation) of this approach is what they are most comfortable with. You will do little to help clients if all you do is sit there and say "Uh-huh." Learn to do many things that can help clients, and then don't be afraid to do them.

Find competence, not in behavior, but in outcomes. Competence does not lie principally in behaviors but in the accomplishments toward which these behaviors are directed. If Brenda has persistently high levels of free-floating anxiety, then one of the hoped-for accomplishments of the helping process is that her anxiety be reduced. If a year later she is still as anxious as ever despite weekly visits to a professional helper, then — other things, such as Brenda's commitment to change, being equal — that helper's competence is in question. Trust in a helper can evaporate quickly if little or nothing is accomplished through the helping sessions.

Respect — Valuing Diversity and Individuality

Respect is such a fundamental notion that it eludes definition. The word *respect* comes from a Latin root that includes the notion of seeing or viewing. Indeed, respect is a particular way of viewing oneself and others. Respect means prizing people simply because they are human. Respect, if it is to make a difference, cannot remain just an attitude or a way of viewing others.

Understanding and valuing diversity.   Helpers differ from their clients in any number of ways — gender, color, sexual orientation, social status, economic status, religion, politics, ethnicity, work experience, age/life stage, type of problem, and so forth. Therefore, as many researchers and practitioners have pointed out (Goodchilds, 1991; Jackson & Meadows, 1991; Lee & Richardson, 1991; Pedersen, 1991a, 1991b; Zane, Sue, Hu, & Kwon, 1991), understanding and valuing diversity is critical to effective helping. A special issue of the *Journal of Counseling and Development* edited by Pedersen (1991c) goes as far as saying that multiculturalism constitutes a fourth force in counseling. Sue, Arredondo, and McDavis (1992) outlined the multicultural attitudes, working knowledge, and skills needed by helpers in three broad areas — helpers' awareness of their own cultural values and biases, helpers' understanding of the world views of their clients, and helpers' ability to use culturally appropriate intervention strategies.

Multiculturalism is part of the people-in-systems approach espoused in Chapter 1. Few practitioners espouse an approach that says "you have to be one to help one" — that is, you must be black to help blacks, you must be male to help males, you must be Asian to help Asians, and on to other forms of diversity. Even so, problems and opportunities are deeply affected by these forms of diversity. Therefore, as we shall see later, a radical empathy needs to pervade helpers' interactions with their clients.

Understanding and valuing the individual.   Although diversity focuses on differences both between and within groups — cultures and subcultures, if you will — helpers interact with clients as individuals, even when counseling is done in small groups. A middle-class black male is *this* individual. A poor Asian woman is *this* person. Any two-way conversation — even one between two identical twins — is, in a very real sense, a *crosscultural* event because each person is a different individual with differences in terms of personal assumptions, beliefs, values, norms, and patterns of behavior. Therefore, an even deeper or more radical empathy is called for in order to address the individuality of clients in that the beliefs, values, norms, and patterns of behavior of their cultures are expressed in each individual's unique way.

Valuing diversity and individuality is not the same as espousing a splintered, antagonistic society or a society of one. Valuing individuals is not the same as promoting radical individualism. Moving to a society of one makes counseling and other forms of human interaction impossible. Individuals are in multiple relationships and all these relationships need to be valued. Here, then, are some norms that apply to human beings, however diverse.

Do no harm. This is the first rule of the physician and the first rule of the helper. Yet some helpers do harm either because they are unprincipled or because they are incompetent. There is no place for the caring incompetent in the helping professions. Helping is not a neutral process; it is for better or for worse. In a world in which such things as child abuse, wife battering, and exploitation of workers are much more common than we care to think, it is important to emphasize a nonmanipulative and nonexploitative approach to clients. Studies show that some instructors do exploit trainees both sexually and in other ways and that some helpers do the same with their clients. This behavior obviously breaches the codes of ethics espoused by all the helping professions.

Appreciate diversity. Respect means prizing those human dimensions that make clients diverse and working especially hard to understand clients different from oneself. A black is not a white person with black skin. A Hispanic Catholic is not the same as an Asian Catholic. They are not just Catholics, nor are they Catholic in the same way. Skilled helpers are sensitive to diversity and they do not let themselves be overwhelmed by it. They see that the problem-management helping process, however adapted, speaks to the underlying humanity and metaphysical makeup of clients. All of us share a common core (Speight, Meyers, Cox, & Highlen, 1991, p. 29).

Treat clients as individuals. Individual differences are as important as cultural and other differences. Respect means prizing the individuality of clients; supporting each client in his or her search for self; and personalizing the helping process to the needs, capabilities, and resources of the client. Skilled helpers do not try to make clients over in their own image and likeness. On the other hand, respect does not mean encouraging clients to develop or maintain a kind of individualism that is either self-destructive or destructive of others. All helping should enable clients to become more effective participants in the social settings of their lives.

Suspend critical judgment. You are there to help clients, not to judge them. Nor are you there to shove your values down their throats. You are there, however, to help them identify, explore, and review the consequences of the values they have adopted. Consider the differences in the following counselors' remarks.

CLIENT: I am really sexually promiscuous. I give in to sexual tendencies whenever they arise and whenever I can find a partner. This has been the story for the past three years at least.

HELPER A: Immature sex hasn't been the answer, has it? Ultimately it is just another way of making yourself miserable. These days, it's just asking for AIDS.

HELPER B: So, letting yourself go sexually is also part of the total picture. You sound as if you're not completely happy about this.

Helper B neither judges nor condones. By picking up the client's affect, she encourages him to explore further. At this stage, she gives the client room to move.

Make it clear that you are for the client. The way you act with clients will tell them how you feel. Your manner should indicate that you care in a down-to-earth, nonsentimental way. Respect is both gracious and tough-minded. Being for the client is not the same as taking the client's side or acting as the client's advocate; it means taking clients' interests seriously even when they need to be challenged. Respect often involves helping clients place demands on themselves. As Fisher and Ury (1981) suggested, "Be hard on the problem, soft on the people" (p. 55). You can challenge clients who are putting themselves and others at risk through promiscuous sexual behavior and still be for them.

Be available. You should be able to say that working with this client is worth your time and energy. Effective helpers are ready to commit themselves to their clients and are available to them in reasonable ways. Just what is reasonable seems to differ from helper to helper. Some helpers don't mind calls in the middle of the night. Other helpers see this as coddling clients to the point of doing them a disservice. Certainly, whenever you are with your clients, you should be all there, you should be psychologically available. This often means putting your own concerns aside for the time being.

Assume the client's goodwill. Work on the assumption that clients want to work at living more effectively, at least until that assumption is proved false. The reluctance and resistance of involuntary clients is not necessarily evidence of ill will. Respect means entering the world of the clients in order to understand their reluctance and a willingness to help clients work through it. The best helpers ask themselves whether they are contributing in any way to the failure of the counseling relationship. Some counselors too easily abandon unmotivated clients. On the other hand, respect does not call for continuing a relationship that is going nowhere.

Be warm within reason. When helpers seek help for themselves, they look for competence, experience, good reputation, and warmth in their counselors (*see* Norcross, Strausser, & Faltus, 1988). Gazda (1973) described warmth as the physical expression of understanding and caring. As such, it is ordinarily communicated nonverbally through gestures, posture, tone of voice, touch, and facial expression. Warmth, however, is only one way of showing respect, and not necessarily the best way. Individual clients differ in their expectations and needs with respect to warmth. Effective counselors, without being phony, gear their expressions of warmth to the needs of the client and not to their own need to either express or withhold warmth. Further, there are limits to warmth. Helpers should be friendly, recognizing that the warmth that characterizes close friendships is not the same as the facilitative warmth of the helping relationships.

Keep the client's agenda in focus. We live in an era during which customer service is getting a great deal of attention. Your clients are your customers and they deserve the best you can give. In particular, helpers should pursue their clients' agendas, not their own. One helper recalled, painfully, an early attempt at helping. His client was a young man who was being beaten up by his lover. This helper, focusing more on the theories he had learned in graduate school than the client's pain, talked to the client about what he considered the client's real problem, the dynamics of the client's sexual orientation. The helper's elation over his success in pinpointing the dynamics correctly turned sour when the young man did not return. In a word, the helper's approach was not pragmatic. It did not work. Appealing to what matters to the client, at least initially, constitutes part of client-centered pragmatism.

Help clients through the pain. Clients might well find the helping process or parts of it painful. Counselors show respect by helping clients through their pain, not by helping them find ways to avoid it. That is, respect includes an assumption on the part of the counselor that the client, right from the beginning, is willing to pay the price of living more effectively. Respect, then, places a demand on the client at the same time that it offers help to fulfill the demand.

Genuineness — Beyond Professionalism and Phoniness

Like respect, helper genuineness refers to both a set of attitudes and a set of counselor behaviors. Some writers refer to genuineness as congruence. Genuine people are at home with themselves and therefore can comfortably be themselves in all their interactions. Being genuine means doing some things and not doing others.

Do not overemphasize the helping role. Genuine helpers do not take refuge in the role of counselor. Ideally, relating at deeper levels to others and helping are part of their lifestyle, not roles they put on or take off at will. This keeps them far away from being patronizing and condescending. Gibb (1968, 1978) suggested ways of being role-free. He said that helpers should learn how to:

- express directly to another whatever they are presently experiencing,
- communicate without distorting their own messages,
- listen to others without distorting the messages they hear,
- reveal their true motivation in the process of communicating their messages,
- be spontaneous and free in their communications with others rather than use habitual and planned strategies,
- respond immediately to another's need or state instead of waiting for the right time or giving themselves enough time to come up with the right response,
- manifest their vulnerabilities and, in general, the mix of their inner lives,
- live in and communicate about the here and now,

- strive for interdependence rather than dependence or counterdependence in their relationships with their clients,
- learn how to enjoy psychological closeness,
- be concrete in their communications, and
- be willing to commit themselves to others.

By this Gibb did not mean that helpers should be free spirits, inflicting themselves on others. Indeed, such helpers can even be dangerous. Being role-free is not license. Freedom from role means that counselors should not use the role or facade of counselor to protect themselves, to substitute for competence, or to fool the client in other ways.

Be spontaneous. Many of the behaviors suggested by Gibb are ways of being spontaneous. Effective helpers are tactful as part of their respect for others but they do not constantly weigh what they say to clients. They do not put a number of filters between their inner lives and what they express to others. On the other hand, being genuine does not require verbalizing every thought to the client.

Avoid defensiveness. Genuine helpers are nondefensive. They know their own strengths and deficits and are presumably trying to live mature, meaningful lives. When clients express negative attitudes toward them, they examine the behavior that might cause the clients to think negatively, try to understand the clients' points of view, and continue to work with them. Consider the following example:

CLIENT: I don't think I'm really getting anything out of these sessions at all. I still feel drained all the time. Why should I waste my time coming here?

HELPER A: If you were honest with yourself, you'd see that you are the one wasting time. Change is hard and you keep putting it off.

HELPER B: Well, that's your decision.

HELPER C: So from where you're sitting, there's no payoff for being here. Just a lot of dreary work and nothing to show for it.

Counselors A and B are both defensive. Counselor C centers on the experience of the client, with a view to eventually helping her explore her responsibility for making the helping process work. Since genuine helpers are at home with themselves, they can allow themselves to examine negative criticism honestly. Counselor C, for instance, would be the most likely of the three to ask himself or herself whether he or she is contributing to the apparent stalemate.

Be open. Genuine helpers are capable of deeper levels of self-disclosure even within the helping relationship. They do not see self-disclosure as an end in itself, but they feel free to reveal themselves, even in deeper ways, when and if it is appropriate. Being open also means that the helper has no hidden agendas; instead, "What you see is what you get."

Client Self-Responsibility — Nonpatronizing Empowerment

Client self-responsibility is a core value of the helping process. Miller (*see* Bevan, 1982) suggested that the "social responsibilities of psychologists can best be discharged by learning how to help people help themselves" (p. 1316). It is assumed, then, that within limits, women and men are capable of making choices and, to some degree, controlling their destinies. For years many psychologists took a deterministic stance and considered the issue of freedom versus determinism a non-question, ignoring the fact that concepts such as self-control and self-determination have a long history in Western thought. A whole range of thinkers from different ideological backgrounds, from Aristotle to Karl Marx, agree that happiness includes the self-respect that comes from accepting responsibility for one's own life and earning one's way in the world.

Studies have shown that self-administered treatment programs are better than no program at all and may often enough be as good as helper-administered treatment programs (*see* Scogin, Bynum, Stephens, & Calhoon, 1990). Counseling, then, should help clients discover and use their power at the service of problem management and opportunity development. McWhirter (1991) has provided us with the following definition of empowerment in counseling:

> Empowerment is the process by which people, organizations, or groups who are powerless (a) become aware of the power dynamics at work in their life context, (b) develop the skills and capacity for gaining some reasonable control over their lives, (c) exercise this control without infringing on the rights of others, and (d) support the empowerment of others in their community. (p. 224)

There is no doubt that social, political, economic, and cultural forces can limit people's ability to exercise self-responsibility. Sometimes counseling others involves helping them exercise self-responsibility in settings in which the odds are overwhelmingly against them. Sometimes counseling involves challenging clients to develop a sense of self-responsibility in a society that unwittingly rewards dependency and helplessness. Helping those who have few incentives to exercise self-responsibility is not easy. Sometimes clients are their own worst enemies, ignoring or wasting whatever power they do have.

Farrelly and Brandsma (1974) outlined a number of hypotheses about client self-responsibility that are translated here into the pragmatics of self-responsibility.

Start with the premise that clients can change if they choose. We seem to be turning into a society of victims, so much so that the plight of real victims, such as children who are victimized by parents, is trivialized. Even if circumstances have diminished a client's degree of freedom, work with the freedom that is left. This is also true of real victims. Even clients with severe problems in living can make headway.

Help clients see counseling sessions as work sessions. Helping, as presented here, is not about exploring the mysteries of life. Rather helping is about client-enhancing change—changing self-defeating beliefs, values, and norms and changing the self-defeating patterns of behavior they spawn. Therefore, counseling sessions deal with exploring the need for change, the kind of change needed, creating programs of constructive change, engaging in change "pilot projects," and finding ways of dealing with obstacles to change. This is work, pure and simple. It can be arduous, even agonizing, but it can also be fun, even exhilarating. Helping clients develop the work ethic that makes them partners in the helping process can be one of the helper's most formidable challenges.

Help clients discover and use their own resources. Clients have more resources for managing problems in living and developing opportunities than they or many helpers assume. The helper's basic attitude should be that clients have the resources to participate collaboratively in the helping process and to manage their lives more effectively. These resources may be blocked in a variety of ways or simply unused. The counselor's job is to help clients free and cultivate these resources. They also help clients assess their resources realistically so that aspirations do not outstrip resources.

Do not overrate the psychological fragility of clients. Neither pampering nor brutalizing clients serves their best interests. However, many clients are less fragile than helpers make them out to be. Helpers who constantly see clients as fragile may well be operating in a self-protective model. Helpers must not be brutal or indifferent, but "tough love" does have its place in helping. What this looks like in practice depends on the state and needs of each individual client. The fact that some clients take so long to act may be due to timidity—the timidity of helper and client alike. Driscoll (1984) noted that too many helpers shy away from doing much more than listening early on. They are fearful of making some kind of irretrievable error if they don't go slowly. Some caution is appropriate, but one can easily become overly cautious. He suggested that helpers intervene more—and right from the beginning; he maintained that most clients appreciate early interventions.

Help clients turn self-dissatisfaction into a lever for change. If clients are dissatisfied with themselves because they should be, your job is not to provide comfort but to help them leverage their dissatisfaction. Even when self-dissatisfaction without any real basis is itself the problem, the client's discomfort can be a lever.

HELPING AS A SOCIAL-INFLUENCE PROCESS

Helping is a social-influence process even though it still is not clear how it works (Dorn, 1986; Heppner & Claiborn, 1989; Heppner & Frazier, 1992; McNeill & Stoltenberg, 1989; Tracey, 1991). As such it is a two-edged sword.

It is not possible for helpers to avoid influencing their clients (or being influenced by them), any more than it is possible to eliminate social influence as a part of everyday life. We are constantly influencing one another in many different ways. Driscoll (1984) put it well when he said that the "obvious objective of [helping] is not merely to understand, but to benefit troubled persons. The emphasis is thus on influence, and on the concepts, understanding, procedures, and competencies used to generate . . . changes" (p. 5). But the main point is that helpers can influence clients without robbing them of self-responsibility.

A Brief History of Social Influence in Helping

The history of social influence in the helping professions is touched with magic. Many cultures have stories about how people suffering from a variety of emotional disturbances and a variety of physical ailments of psychogenic origin have been cured by their belief in the curing powers of a helper (Frank, 1973; Kottler, 1986). Very often such cures have taken place in religious contexts, but they have not been limited to such contexts. In the typical case, people come to see a certain person as being a healer with great powers. This person might be a tribal shaman or a Western helper with a good reputation. Prospective clients hear that such healers have cured others with ailments similar to theirs. They generally see these healers as acting, not in their own interests, but in the interests of the afflicted who come to them. This belief enables the afflicted person to place a great deal of trust in the healer. Finally, in a ceremony that is often public and highly emotional, the healer in some way touches the afflicted person either physically or metaphorically, and the person is "healed." The tremendous need of the afflicted person, the reputation of the healer, and the afflicted person's trusting belief in the healer all heighten the person's belief that he or she will be cured. In fact, in cases where sufferers are not cured, the failure is often attributed to their lack of belief or to some other evil within them (for instance, possession by a demon or poor motivation). Then they not only remain with their afflictions but also lose face in the community, becoming outcasts or crazies.

The dynamics of such cures are hard to explain empirically. It is obvious that elements of the healing process help marshal the emotional energies and other resources of the afflicted. For instance, sufferers experience hope and other positive emotions, which they perhaps have not experienced for years. The whole situation mobilizes their resources and places a demand on them to be cured. It presents them with what the Greeks called *kairos*—an opportune, acceptable, favorable, legitimate moment in time to leave their old way of life behind and take up a new one. The power of suggestion in such cases can be great, even overwhelming. Skilled helpers are aware of and have a deep respect for such positive shadow-side factors in the helping process.

In 1968 Strong wrote what proved to be a landmark article on counseling as an interpersonal-influence process. The article is considered landmark because it has stimulated a great deal of writing and research that continues to this day. Strong emphasized the perceived expertise, attractiveness, and trustworthiness of helpers as the principal sources of influence. Strong went on to suggest a two-stage process in helping. First, counselors enhance their perceived expertness, attractiveness, and trustworthiness in order to get their clients involved in the helping process. Second, helpers use their influence to help clients change self-defeating beliefs, values, and norms together with self-defeating patterns of behavior. In other words, helpers establish a power base and then use this power to influence clients to do whatever is necessary to manage their lives more effectively. Most research so far has focused on the first stage of Strong's model.

Goldstein (1980) described the social-influence process in helping in a way that was complementary to Strong's. He calls behaviors by which helpers establish an interpersonal power base with their clients "relationship enhancers." In essence, these enhancers are the values we have just reviewed, put into practice.

Reconciling Self-Responsibility and Social Influence

It is especially important for helpers to come to grips with helping as a process of social influence in view of the unilateral nature of helping. Imagine a continuum. At one end lies telling clients what to do and at the other leaving clients to their own devices. Somewhere along that continuum is helping clients make their own decisions and act on them. Preventing a client from jumping off a bridge moves to the controlling end of the continuum; accepting a client's decision to leave therapy even though he or she is "not ready" moves toward the other end. Most forms of helper influence fall somewhere in between. As Hare-Mustin and Marecek (1986) noted, there is a tension between the right of clients to determine their own interests and the therapist's obligation to protect the client's welfare.

Here are some of the things you can do to protect the client's self-responsibility.

Use a participative rather than a directive model of helping. We have already discussed helping as a collaborative effort on the part of helper and client. The model spelled out in these pages constitutes, at its best, a shared process. Continually search out ways to get clients to own their share of the process. Counseling provides a transitory support system. Increased self-responsibility is the objective of this support.

Help clients discover and use the power they have. Today there is a great deal of talk about empowerment. We empower employees and we empower clients. In many cases, however, the power is already there. We merely help clients use it. In a classic work, Freire (1970) warned helpers against making helping itself just one more form of oppression for the already oppressed. He helped victims of government oppression find, develop, and use their own power.

Accept helping as a natural, two-way influence process. Helping is a two-way street. Clients and therapists change one another in the helping process. Even a cursory glance at helping reveals that clients can affect helpers in many ways. Helpers find clients attractive or unattractive and must deal with both positive and negative feelings about them as well as manage their own behaviors. They may have to fight the tendency to be less demanding of attractive clients or not to listen carefully to unattractive clients. On the other hand, some clients trip over their own distorted views of their helpers. A young woman who has had serious problems in her relationship with her mother might begin to have problems relating to a helper who is an older woman. Unskilled helpers can get caught up in their own and in their clients' games. Skilled helpers understand the shadow side of the helping relationship and manage it.

Become a consultant to clients. Helpers can see themselves as consultants hired by clients to help them face problems in living more effectively. Consultants in the business world adopt a variety of roles. They listen, observe, collect data, report observations, teach, train, coach, provide support, challenge, advise, offer suggestions, and even become advocates for certain positions. But the responsibility for running the business remains with those who hire the consultant. Therefore, even though some of the activities of the consultant can be seen as quite directive, the decisions are still made by managers. Consulting, then, is a social-influence process, but it is a collaborative one that does not rob managers of the responsibilities that belong to them. In this respect it is a useful analogy to helping. The best clients, like the best managers, learn how to use their consultants to add value.

Democratize the helping process. Tyler, Pargament, and Gatz (1983) moved a step beyond consultancy in what they called the "resource collaborator role." Seeing both helper and client as people with defects, they focused on the give-and-take that should characterize the helping process. In their view, either client or helper can approach the other to originate the helping process. Both have equal status in defining the terms of the relationship, in originating actions within it, and in evaluating outcomes and the relationship itself. In the best case, positive change occurs in both parties.

If social influence means pushing clients to accept the counselor's point of view, then clients will inevitably feel coerced and do what we all tend to do when coerced — resist actively or passively. Even worse, some clients might mindlessly accept what we say, even when it is not in their self-interest. Social influence at its best means that clients will listen to and weigh what helpers say to determine whether it will help them manage their lives better. Finally, work has been done on how clients themselves can use social influence processes to control and manage problems in their everyday lives (see Edwards, Tindale, Heath, & Posavac, 1990).

THE CLIENT-HELPER
WORKING CHARTER

Explicit and implicit contracts govern the transactions that take place between people in a wide variety of situations, including marriage (where
some, but by no means all, of the provisions of the contract are explicit) and
friendship (where the provisions are usually implicit). That is, there is an
extensive shadow side to both explicit and implicit contracts. In counseling,
the client-helper contract has been, traditionally, implicit, even though the
need for more explicit structure has been discussed for years (Proctor &
Rosen, 1983). Because of this, the expectations of clients may differ from the
expectations of their helpers (Benbenishty & Schul, 1987). That is, clients
have been expected to accept what the helper has to offer without necessarily understanding what it was. Implicit contracts are not enough. More
and more professionals are urging helpers to discuss the fundamentals of
the contract being struck (Woody, 1991). Some say that discussions are not
enough and that the contract should be in writing (Handelsman & Galvin,
1988; Weinrach, 1989). If helping is to be a collaborative venture, then both
parties must understand what their responsibilities are. Perhaps the term
working charter is better than contract. It avoids the legal implications of the
latter term and connotes a cooperative venture.

     The working charter need not be too detailed, nor should it be rigid.
The question is: how much structure will help this client at this time? The
charter needs to provide structure for the relationship and the work to be
done without frightening or overwhelming the client. Ideally, the working
charter is an instrument that makes clients more informed about the process, more collaborative with their helpers, and more proactive in managing
their problems. At its best, a working charter can help client and helper
develop realistic mutual expectations, give clients a flavor of the mechanics
of the helping process, diminish initial client anxiety and reluctance, provide a sense of direction, and enhance clients' freedom of choice. To achieve
these objectives the working charter should include, generically, an overview of the helping process, the techniques to be used, and its flexibility;
how this process will help clients in the long run; how the relationship is to
be structured and the kinds of responsibilities clients and helpers will have;
and procedural issues.

The helping process.   Helping should not be a black box whose internal
workings are unknown. Clients have a right to know what they are getting
into. A simple handout, whether graphic, text, or both, can serve as the
starting point. Just what kind of detail will help differs from client to client.
Obviously clients should not be overwhelmed by distracting detail from the
beginning. Nor should highly distressed clients be told to contain their anxiety until helpers teach them the helping model. Rather the details of the
model can be shared over a number of sessions. A simple pamphlet outlin-

ing the stages and steps of the helping process can be of great help, provided that it is in language that clients can readily understand. In my opinion, clients should be told as much about the model as they can assimilate. Like helpers, clients can use the model as a cognitive map to give themselves a sense of direction. If clients are to be collaborators, they need the map.

Counseling should be presented as a flexible, client-serving process rather than a rigid, model-serving process: "I have outlined what the helping process might look like. But, in the end, everything we do here should help you manage your problems better. There are no rigid rules about the helping process. We can do anything that is ethical and useful." This reinforces the pragmatism of helping. Explaining the counseling process does not lock clients and helpers into a single way of doing things. Flexibility at the service of the client is a value. Effective helpers are natural, and it is only natural to change things when they are not working. An effective helper might say, "If what we're doing together is not working, then we can change it. This is all about helping you manage your concerns better."

The client as problem manager.   If the second goal of the helping process is to be achieved — that is, if clients are to go away better able to manage their problems in living more effectively on their own — then sharing some form of the problem-management process with them is essential. In an ideal world, clients would have learned problem solving when they were young (Elias & Clabby, 1992). But this is the exception rather than the rule. Therefore, many people grow up doubting their own ability to manage problems in any kind of systematic way even though belief in one's ability to solve problems is itself a strength (Nezu, 1985). Since the problem-management process is one of the principal tools that clients can use to manage their problems in living more effectively, during and after the formal helping sessions, helping them learn the process is of paramount importance. Edward Deming, a quality guru, made an interesting point during a television program dealing with quality in the workplace in 1992. He was asked: "Don't we get better by doing things? Isn't experience one of our best teachers?" His response would surprise many. He said that experience teaches us nothing *unless we have some sort of theory or model or framework that helps us make sense of our experience and apply it to other areas of our lives.* The helping process, since it is also a self-help process, is such a framework. It enables clients to learn from experience both within the helping sessions and during their struggles to implement a program for constructive change in their day-to-day lives.

The relationship.   The working charter should help structure the relationship, perhaps as "I'd like to become a partner with you in helping you manage your concerns more effectively." The value package you choose to work from should find its way into a discussion about what the relationship will look like. Again, be careful not to overwhelm clients with unneeded detail from the beginning. An explanation of the relationship should include the responsibilities of the helper: "I want to make sure that I understand your concerns. If I do that, then you will come to understand them better and be in a better position to do something about them." Or, "I'm not going to solve your problems for you — that doesn't work — but I am going to do my best to help you in your search for ways of managing your concerns better." If client self-responsibility is important, then it is essential to discuss at some time and in some way, the client's responsibilities within the helping sessions and in the world. "In the end, counseling is not about talking, but about acting. If people are to manage their lives better, they usually have to act differently. I'd like to help you do that, but, of course, I can't do it for you." Finally, since helping is a social-influence process, how this will work should be discussed. "If I see you trying to avoid doing something that is for your own good, I'll tell you about it. Or rather I will invite you to challenge yourself. But I will not try to force you to do anything."

Procedural issues.   The nuts and bolts of the helping process — where sessions will be held, how long they will last, and so on — should be communicated. Procedural limitations (for instance, how free the client is to contact the helper between sessions) should also be discussed. "Ordinarily we won't contact each other between sessions, unless we prearrange it for a particular purpose. However, . . ." Essential ground rules should not come as a surprise to clients. Moreover, the flexibility of these rules should, at least eventually, become clear.

Obviously, you may not use this specific list or the explanations provided. But, just as it is essential that you review the values that will drive the helping process, you need to determine what you are to tell clients about the process itself. Different clients have different needs, and what you share with them should be geared to their ability to assimilate and use the information. Some clients might be helped by a great deal of information up front. Others might appreciate getting information in stages. For them an outline up front is enough; they can learn the details as they go along. In short, the working charter between you and your clients needs to be explicit and clear, but there are many ways of accomplishing this.

THE SHADOW SIDE OF VALUES

At their best, values are not just beliefs, but pragmatic criteria we use to drive behavior and make decisions. For instance, if wellness is a value, it becomes a criterion for choosing an approach to nutrition and exercise. Values, then, are essential guidelines for structuring the client-helper relationship and spelling out the key features of the helping process. We need only to look at our own experience to understand the shadow side of values:

- We do not always have a clear idea of what our values are.
- The values we say we believe in, our espoused values, do not always coincide with our actions, our values-in-use. That is, values remain good ideas, not drivers of behavior.
- Expediency has us compromise our values even though we are very good at coming up with reasons why the compromises were necessary.

Some helpers fail to outline and debate for themselves the values that are to drive the helping process. Others find themselves taking shortcuts. For instance, even though they hold that they should not make decisions for clients, they do so anyway out of frustration.

When it comes to sharing the helping process itself, some counselors are reluctant to let the client know what the process is all about. Of course, helpers who fly by the seat of their pants can't tell clients what it's all about since they don't know what it's all about themselves. Still others seem to think that knowledge of helping processes is secret or sacred or dangerous and should not be communicated to the client. Less savory are situations in which bait-and-switch tactics are used to influence clients (M. H. Williams, 1985). Williams discussed the case of a man who had been persuaded to see a therapist because of his hypertension. In therapy, however, the helper wanted to focus on the man's sex life.

CHAPTER 4

A BIAS TOWARD ACTION

THE IMPORTANCE OF A BIAS
TOWARD ACTION

Shakespeare, in the person of Hamlet, talks about important enterprises, what he calls "enterprises of great pith and moment," losing "the name of action." Helping, which is certainly an enterprise "of great pith and moment," can lose the name of action. One of the principal reasons clients do not manage the problem situations of their lives effectively is their failure to *act* intelligently and prudently in their own best interests. Covey (1989), in his immensely popular *The Seven Habits of Highly Effective People*, named proactivity as the first habit: "It means more than merely taking initiative. It means that as human beings, we are responsible for our own lives. Our behavior is a function of our decisions, not our conditions. We can subordinate feelings to values. We have the . . . responsibility to make things happen" (p. 71). Counselors fail their clients by not helping them become proactive. Helping becomes a process of too much talking and too little action. Many approaches to helping discuss the need for clients to act and suggest ways of helping them do so, but too often action is tucked away at the end of the model, after all the talking is over — as if it were an afterthought — or forgotten altogether.

A counselor at a large manufacturing concern realized that inactivity did not benefit injured workers. If they stayed at home, they tended to sit around, gain weight, lose muscle tone, and suffer from a range of psychological symptoms such as psychosomatic complaints unrelated to their injuries. Taking a people-in-systems approach, she worked with management, the unions, and doctors, to design temporary, physically light jobs for injured workers. In some cases nurses or physical therapists visited these workers on the job. Counseling sessions helped to get the right worker into the right job. The workers, active again, felt better about themselves and the company benefitted.

Harvard's Argyris (1957, 1962, 1964) outlined a range of tasks to be accomplished in moving toward adult maturity. All of them require, directly or indirectly, a bias toward action in people's lives:

- Assume an active rather than a passive role in life.
- Change from a state of dependency on others to relative independence.
- Widen your range of behaviors. Act in many rather than few ways.
- Develop a wider range of interests — moving from erratic, shallow, and casual interests to mature, strong, and enduring interests.
- Change from a present-oriented time perspective to a perspective encompassing past, present, and future.
- Move from solely subordinate relationships with others to relationships as equals or superiors.
- Change from merely understanding yourself to some kind of control over your destiny.

Helpers see many clients who have difficulties in one or more of these areas, clients who need assistance in becoming agents in their own lives. If we overemphasize the ways in which personal agency can be limited by environmental, biological, social, and emotional factors, then we risk turning all clients into victims (*see* Howard & Conway, 1986; Howard, Curtin, & Johnson, 1991; Howard & Myers, 1990).

EXPERIENCES, BEHAVIORS, AND AFFECT

Traditionally, human activity is divided into three parts — thinking, feeling, and acting. A slightly different approach is taken here to reflect what clients talk about in their dialogues with helpers.

- Clients talk about their *experiences* — that is, what happens to them. If a client tells you that she was fired from her job, she is talking about her problem situation in terms of an experience.
- Clients talk about their *behavior* — that is, what they do or refrain from doing. If a client tells you that he has sex with underage boys, he is talking about his problem situation in terms of his behavior.
- Clients talk about their *affect* — that is, the feelings and emotions that arise from or are associated with either experiences or behavior. If a client tells you how depressed she gets after drinking bouts, she is talking about the affect associated with her problem situation.

I usually find that clients are fairly willing to talk about their experiences, less willing to talk about their feelings and emotions, and least willing to talk about their actions.

All three, of course, are totally interrelated in the day-to-day lives of clients. In counseling dialogues, then, clients talk about all three together. Consider this example. A client says to a counselor in the personnel department of a large company: "I had one of the lousiest days of my life yesterday." At this point the counselor knows that something went wrong and that the client feels bad about it, but she knows relatively little about the specific experiences, behaviors, and feelings that made the day a horror for the client. However, the client continues: "Toward the end of the day my boss yelled at me for not getting my work done [an experience]. I lost my temper [emotion] and yelled right back at him [behavior]. He blew up and fired me [an experience for the client]. And now I feel awful [emotion] and am trying to find out [behavior] if I can get my job back." Note that the problem situation is much clearer once it is spelled out in terms of specific experiences, behaviors, and feelings related to specific situations.

Experiences.   Most clients spend a fair amount of time, perhaps too much time, talking about what happens *to* them.

- "I get headaches a lot."
- "My ulcers act up when family members argue."
- "My wife doesn't understand me."

This kind of talk often focuses on what other people do or fail to do. The implication is that others are to blame for any problems:

- "She doesn't do anything all day. The house is always a mess when I come home from work."
- "He tells his little jokes, and I'm always the butt of them."
- "She never calls."

However, the controlling forces may be internal or from unknown sources:

- "These feelings of depression come from nowhere and seem to suffocate me."
- "I always feel hungry."
- "I just can't stop thinking of her."

One reason that some clients are clients is that they see themselves as victims, adversely affected by other people, the immediate social settings of life, society in its larger organizations and institutions, cultural prescriptions, or even internal forces. They feel that they are no longer in control of their lives or some dimension of life. Therefore, they talk extensively about these experiences.

- "Company policy discriminates against women."
- "The economy is so lousy that there are no jobs."
- "No innovative teacher gets very far around here."

For some clients, talking constantly about experiences is a way of avoiding responsibility: "It's not my fault. After all, these things are happening to me." In *A Nation of Victims*, Sykes (1992) is troubled by what he sees as the tendency of the United States to become a "nation of whiners unwilling to take responsibility for our actions." A school-district employee fired for chronic tardiness can sue for reinstatement because chronic tardiness can be seen as a disability. Sykes says that just as lawyers tell us there is always somebody to sue, psychotherapists teach us that life should be pain-free. That is certainly not the value stance here. Nor do I think that it is the common value stance in the helping professions.

Behaviors.   All of us do things that get us into trouble and fail to do things that will help us get out of trouble. Clients are no different.

- "When he ignores me, I think of ways of getting back at him."
- "I haven't even begun to look for a job. I know there are none in this city."

- "Even though I feel the depression coming on, I don't take the pills the doctor gave me."
- "When I get bored, I find some friends and go get drunk."
- "Not only am I promiscuous, but I engage in unprotected sex whenever my partner will let me."

The reason that some clients talk freely about their experiences, but are much more reluctant to talk about their behaviors is simple. While we may not feel accountable for our experiences, what happens to us, we realize at some level of our being that we are responsible for what we do or actively refrain from doing. Talking about behavior places a demand on clients to consider the possibility of acting differently.

Affect.   Affect refers to the feelings and emotions that proceed from, lead to, accompany, underlie, or give color to a client's experiences and behaviors.

- "I finished the term paper that I've been putting off for weeks and I feel great!"
- "I've been feeling sorry for myself ever since he left me."
- "I yelled at my mother last night and now I feel pretty ashamed of myself."
- "I've been anxious for the past few weeks, but I don't know why. I wake up feeling scared and then it goes away but comes back again several times during the day."

Of course, clients often express feelings without talking about them. Some clients feel deeply about things, but do their best to repress their feelings. But often there are cues or hints, whether verbal or nonverbal, of the feelings inside. A client who is talking listlessly and staring down at the floor may not say, in so many words, "I feel depressed." A dying person might express feelings of anger and depression without talking about them.

Some clients imply that their emotions have a life of their own and that they can do little or nothing to control them. This includes describing others as the causes of their emotions:

- "Whenever I see him with her, I feel hurt."
- "She can get my goat whenever she wants. She's always making me angry."
- "I can't help crying when I think of what they did to me."

One goal of the helping process is to help clients get out from under the burden of disabling feelings and emotions. People can learn how to control their emotions without becoming lifeless. Usually feelings and emotions cannot be controlled directly. The best way to control emotions is to control the thoughts that provoke them. If I let myself dwell on the ways in which you have wronged me, I will inevitably get angry. But I can actively refrain from dwelling on such thoughts. On the other hand, some clients need to learn how to express emotions as part of their humanity and as a way of enriching their interactions with themselves and others.

THE MANY FACES OF ACTION

Counselors need to help their clients do two things related to action: talk about their problem situations and unused opportunities in terms of action, what they do and don't do; and initiate problem-managing action. There are five ways of categorizing behavior or action in the helping process: internal versus external actions, within-sessions versus real-life actions, informal versus formal actions, influencing versus accommodating actions, and actions that lead to positive versus actions that lead to negative outcomes. It will become clear that these categories overlap and interact with one another.

Internal versus External Action

Actions can be internal—inside the head of the client—or external—actions that can be seen by others. Both are central to the helping process.

Internal actions. Many cognitive processes, those things that go on inside our heads, are really forms of internal behavior. These include such actions as thinking, expecting, attending to things inside or outside oneself, registering information, daydreaming, rehearsing responses to others, turning things over in one's mind, making associations, imagining, thinking about one's inner life, recalling, planning, deciding, and permitting thoughts to hang around. Some of these happen without any agency on the part of the client. Others are often deliberate; that is, they are internal actions. Some statements made by clients in telling their stories follow. Note that they describe internal actions.

- "When he called me a name, I began thinking of the ways I could get back at him."
- "I like to daydream about having a child."
- "I never let myself think bad things about another person."
- "When she left me, I decided to forget about her completely."
- "I try not to let thoughts of sex enter my mind."

People who feel that they have little or no control over what is going on inside their heads often describe internal actions as experiences:

- "I try to get into my work, but I just can't stop thinking of her and how she played me for a fool."

- "Sexual fantasies seem to keep popping up all the time. I just don't seem to have any control."
- "I know I get depressed and cry when I think of John on his deathbed, but I can't help thinking of him."

Attitudes and prejudices, as ingrained ways of thinking about things, are more like experiences than actions, but — as we know well — they can lead to powerful and sometimes destructive actions. Furthermore, people reinforce attitudes and prejudices by dwelling on them.

One of the goals of counseling is to help clients see that they can control their internal actions — and experiences — more than they first think. For instance, clients can work at changing attitudes and prejudices. Indeed, some forms of therapy are based on primarily learning how to control what goes on inside one's head. One client, Ben, a man who had lost his wife and daughter in a train wreck, discovered that he was allowing himself — almost encouraging himself — to dwell morbidly on their deaths. He dwelled on thoughts that made him feel guilty. Learning how to refrain from such thoughts became one of the goals of counseling. His counselor helped him find ways of remembering and mourning his wife and daughter that uplifted his spirit instead of destroying it. He discovered that he had not really mourned their deaths. Instead, he had indulged in guilt and self-recrimination. He ultimately made a pact with himself to eliminate morbid, self-recriminatory thinking.

External actions.   External action (overt behavior) is action that can be witnessed by others, whether it is actually witnessed or not. Here are some examples of overt behavior in the stories clients have told:

- "When he called me a name, I punched him."
- "I haven't told my husband about the affair."
- "I go to pornographic movie theaters a lot."

Helping is not a head trip. Ordinarily, internal actions on the part of clients — thinking the right things, if you will — are not enough to effect change that makes a difference. Internal action prepares the way for external action. Maria, a cancer patient, worked hard, as she said, to "get her head in order." However, at first she failed to complement this internal work with external actions. With the help of a counselor she began doing things. She started a fitness program. She figured that the healthier her body, the better would she be able to cope with cancer. She also found that her problems always seemed "not as bad" after vigorous exercise. She applied the same thinking to what she called her "stuck on me and my problems" attitude. She began to do some volunteer work. In doing so, she became less preoccupied with her own problems. All in all, selected external actions led to a much more satisfying life.

Actions within the Helping Sessions
versus Actions Outside

Actions can take place during the helping sessions themselves or outside the sessions in the client's day-to-day life.

Inside helping sessions.   Clients must actively participate in and own the helping process. It cannot remain something that is done to them, something to which they merely submit. Talking within the session can be a form of outcome-producing action. Clients who have kept problems to themselves for years can make significant breakthroughs just by talking about them. Many clients have changed their negative views of themselves and made improvements in their interpersonal styles through their interactions with their helpers. One of my clients significantly altered his bossy, confrontational style by first changing his style with me. In most cases, however, the ultimate fruits of social-emotional reeducation lie in changed behavior outside the helping sessions. At its best, both cognitive and behavioral change inside helping sessions prompts change.

Crystal had been a victim of torture in a totalitarian state. At first she took very little initiative in the counseling sessions. Later, moved by her counselor's acceptance and sensitivity, she became more active. She even allowed herself to cry, something she had not done since the torture. Gradually, she began bringing bits and pieces of her own agenda to the helping sessions. Although basically attractive, she saw herself from an emotional point of view as disfigured and ugly. One day she said to the counselor, "Help me explore my ugliness." At first the counselor was surprised, but through the give-and-take of their dialogue Crystal began to see her feelings of "ugliness" slowly subside. Her taking an active stance in the session accelerated the process of social-emotional relearning.

Real-life settings.   I know someone who participated in a counseling group for over 15 years. He looked forward to the meetings and felt bad if he had to miss one. Indeed, the group was an important part of his social life. But there is some question as to how effectively he translated what he learned inside the group to his everyday life. His dissatisfaction with himself, his career, and his social life persisted. It was his life, of course, and he could do what he wanted with it. He sometimes groused about what this was costing him, but this "purchase of friendship" pattern persisted.

What happens within helping sessions should drive behavior outside in the client's real-life settings. An examination of significant between-session behavior and its impact should always be a central part of each session. Andrew was seeing a counselor as part of a court settlement in a sexual harassment case. He had moved from total disbelief to a guarded willingness to "spy on himself" as he interacted with women in everyday settings. Once he stopped being defensive, he came to realize that his actions with women were less than noble. He identified patterns of behavior that were unacceptable — physical closeness, touching, speech peppered with double meanings, the constantly roving eye — and made specific efforts at changing them. He discussed his success and failures with his helper. One payoff was that women began to be more relaxed around him.

Informal versus Formal Actions

Actions may be informal — not part of an overall plan — or formal — part of an overall plan.

Informal actions.   As illustrated in the overview of the helping model, informal action — as opposed to action based on some specific plan — can start right from the first interview. In some cases, more is accomplished through informal action than through formal planning. A friend of mine who is highly successful did not set demanding career goals and then devise a plan for achieving them. Instead, without reflecting on it, he took an incremental approach. He kept saying yes to offers that felt a little bit beyond his current level of expertise. He developed a pattern of stretching himself bit by bit. His approach to his career has been informal and incremental, but it has been action-based and highly effective. Some clients prefer mostly informal actions and goals. Some are helped by formal planning involving explicit goals and action programs. Some benefit from a combination of the two.

Caroline had been brought to the notice of the Family Services Bureau because of suspected abuse of her mentally retarded son; she sought counseling voluntarily. She never discussed any kind of massive change program with her helper, but she began to do little things differently at home. When she came home from work especially tired and frustrated, she had a friend in the apartment building stop by. This helped her from taking her frustrations out on her son. Instead of staying cooped up over the weekend, she found simple things to do that eased tensions, such as going to the zoo and to the art museum. She discovered that her son enjoyed these pastimes immensely even though he didn't understand everything around him. She discovered little ways of becoming his friend and not just his mother. She talked about these "little actions" with the counselor who helped her discover even others. A big breakthrough happened almost by chance. She went to a church carnival with her son and met another woman with a retarded child. She discovered that the church was the location for a self-help group for mothers with difficult children or children with difficulties.

Formal actions.   Actions that are part of a formal plan can be most useful. In rebuilding his life after a serious automobile accident, one of my clients very deliberately planned both a rehabilitation program and a career change. Keeping to a schedule of carefully planned actions not only helped him keep his spirits up but also helped him accomplish a succession of goals. These small triumphs buoyed his spirits and moved him, however slowly, along the rehabilitation path. As already noted, most formal plans don't work because they are never really tried. It is not planning that is the waste of time but planning not followed by action.

Influencing Mode versus Accommodating Mode

There are actions that clients take to influence existing realities and actions they take to accommodate themselves to these realities. Carlos, for instance, persists and is successful in getting the dean to change one of his classes. Weisz, Rothbaum, and Blackburn (1984) call this kind of influence primary control. They call another kind of accommodation secondary control. Candy provides an example of secondary control: she takes the schedule of classes she gets and, even though the hours conflict with other interests, tries to find ways of putting up with it.

Influencing mode.   One of the goals of counseling is to help clients become more effective agents in life; to help them become doers rather than mere reactors, preventors rather than fixers, initiators rather than followers. This is especially the case if the client has become mired down in a problem situation because of a bias against rather than for action. Consider these two cases.

> Nicholas was a single father taking care of an infirm mother. When he finally saw a helper he was depressed because "life had become so narrow." The counseling sessions helped him move from accommodating mode to influencing mode. It became clear that much of his mother's "infirmity" was self-inflicted. He arranged separate living quarters and helped her find a job to supplement the pension income she received from her deceased husband's company. While he remained quite decent, he did not tolerate abuse for "abandoning" her.

> Frank was liked by his superiors for two reasons. First, he was competent — he got things done. Second, he did whatever they wanted him to do. They moved him from job to job when this suited them. He never complained. However, as he matured and began to think more of his future, he realized that there was a great deal of truth in the adage, "If you're not in charge of your own career, no one is." After a session with a career counselor, he outlined the kind of career he wanted and presented it to his superiors. He pointed out to them how this would serve both the company's interests and his own. At first they were nonplussed, but then agreed. Later, when they wanted to break the implicit contract developed in this interchange, he stood his ground.

Nicholas and Frank each made something happen. They did not just suffer in silence. They did not accommodate to the status quo.

Accommodating Mode.   Helping is the art of the possible. While most clients can do more to control their lives than they actually do, this does not mean that clients must substitute taking charge of everything for passivity and dependency. Take the case of Debbie. She learned, to her chagrin, that she and her husband had more areas of incompatibility than she had imagined. Though she knew that he liked his work, she had no idea that it was his passion. For him, work came before almost everything else. Debbie did not think that she could make him change and a counselor helped her find ways of coping. Instead of feeling sorry for herself, she developed friends and interests outside the home, activities in which he did not participate. In her mind, this was not the perfect solution. But it certainly made life fuller and more livable.

Leahey and Wallace (1988) offered the following example of a client in accommodating mode.

> For the last five years, I've thought of myself as a person with low self-esteem and have read self-help books, gone to therapists, and put things off until I felt I had good self-esteem. I just need to get on with my life, and I can do that with excellent self-esteem or poor self-esteem. Treatment isn't really necessary. Being a person with enough self-esteem to handle situations is good enough for me. (p. 216)

The following client also redefines the problem situation, seeing it now from a different angle.

> I would say that I am completely cured. . . . I can still pinpoint these conditions which I had thought to be symptoms. . . . These worries and anxieties make me prepare thoroughly for the daily work I have to do. They prevent me from being careless. They are expressions of the desire to grow and to develop. (Kora, 1967, quoted in Weisz, Rothbaum, & Blackburn, 1984, p. 964)

Some helpers, reviewing these last two examples, would be disappointed. Others would see them as legitimate examples of adapting to, rather than changing, reality. But in both cases, the clients have acted and done something about the way they think about themselves and their problems. For some clients, a great deal of primary control might be possible; for others, secondary control might be the only realistic option; for most, the best available course is a combination of the two. Clients, like ourselves, have limitations. Like ourselves, they are in many ways not captains of their own fates. Some would say that what Debbie did amounted to giving up. Others would see it as a case of accommodation.

Unproductive versus Productive Actions

As Tom Gilbert (1978) so aptly noted, human competence is not found in human behavior. Rather it is found in human accomplishments. Clients often get into trouble because they engage in unproductive activities. To deserve the name of problem-managing or opportunity-developing action, action must lead to client-enhancing outcomes. What difference does it make if the clients leave helping sessions with smiles on their faces and songs in

their hearts if their problems are not managed more effectively? It doesn't —
unless the originating problem was the lack of a smile and the absence of a
song.

Unproductive actions.   Some things clients do actually make things worse.
But even activities that seem to be going somewhere positive may end up
being no more than smoke. Xantha always seemed busy but never seemed
to accomplish much. She quit her job because she wanted to go out on her
own as a consultant. She kept drawing up and revising brochures but never
came out with the final version, she kept looking at office space but never
signed a lease, she kept going to seminars to prepare herself for her work
but never did the marketing to get clients. In sum, even though her life was
filled with activity, she never accomplished anything of substance. Unfortunately, both she and her counselor permitted this pattern to continue during the helping sessions. Voluble as she was, she discussed all these issues
at length, she did all sorts of things — she drew up a marketing plan, she met
with people to determine whether proposed workshops might meet their
needs, she designed and wrote a newsletter for fellow professionals — but
still no actual clients, no actual workshops, no actual dissemination of the
newsletter. The helper, who had been seduced by this flurry of activity,
finally realized he had been had. He ended one session by saying to her,
somewhat unfairly at this point, for he had been a co-conspirator, "Come
back and see me when you have your first real client."
    Early in my career, one of my clients kept a detailed record of every-
thing that happened in the counseling sessions and everything outside that
seemed in any way related. Good idea? Probably not. It took up a great deal
of his time and in the end was probably his way of putting off problem-
managing action. It made him focus excessively on the helping process itself
instead of on behavioral change in his everyday life. He was spinning his
wheels. Now if he had kept a record of problem-managing accomplishments both within and outside sessions, that might have made a difference.

Productive actions.   Actions are productive to the degree that they help the
client, directly or indirectly, move toward problem-managing and
opportunity-developing outcomes. These outcomes can relate to goals to be
achieved within the sessions, goals to be achieved outside the sessions,
short-term goals, intermediate goals, or long-term goals. Lynn came from a
sheltered background. He grew up in a small community where religion
played a large role. When he went to the state university, he began to experiment with different values. He drank a bit, let go a bit, did drugs a bit,
slacked off a bit — but the bits added up and he finally attempted suicide.
During the counseling sessions after his release from the hospital, he focused on "giving things up." His goal was to give up drink, give up loose
living, give up drugs, give up casual sex; that is, he wanted to give up everything he thought had brought him to the brink. The counselor pointed out

that he was creating a vacuum for himself. Lynn knew what he didn't want, but not what he did want. After that the sessions took on a different cast. Lynn talked about developing a social life. What he had been into before was not really a social life. He cautiously joined a photography club and met a few people with whom he felt compatible. He looked around for a church group that was not as rigid as the one back home. He began by attending services and finally volunteered for a few activities that got him back into community. He still felt "different," but these activities put him on the road toward his goal, a reasonable social life.

The possibilities for problem-managing action, then, are limitless. The right action mix will differ from client to client. Some clients need more internal work; others need to get acting immediately in their day-to-day worlds. There are no ready-made formulas. Clients, in consultation with their helpers, need to create their own.

THE SHADOW SIDE OF DISCRETIONARY CHANGE: THE MANY FACES OF INERTIA

Early in the history of modern psychology, William James remarked that few people bring to bear more than about 10% of their human potential on the problems and challenges of living. Others since James, while changing the percentages somewhat, have said substantially the same thing, and few have challenged their statements (Maslow, 1968). It is probably not an exaggeration to say that unused human potential constitutes a more serious social problem than emotional disorders, since it is more widespread. Maslow (1968) suggests that what is usually called normal in psychology "is really a psychopathology of the average, so undramatic and so widely spread that we don't even notice it ordinarily" (p. 16). Many clients you will see, besides having more or less serious problems in living, will also probably be chronic victims of self-inflicted psychopathology of the average.

Discretionary versus Nondiscretionary Change

The distinction between discretionary and nondiscretionary change is critical for helpers. Nondiscretionary change is mandated. If the courts were to say to a divorced man negotiating visiting rights with his children, "You can't have visiting rights unless you stop drinking," then the change is nondiscretionary. There will be no visiting rights without the change. In contrast, a man and wife having difficulties with their marriage are not under the gun to change the current pattern. This change is discretionary. The shadow-side principle here is quite challenging: *In both individual and organizational affairs, the track record for discretionary change is very poor.* This fact is central to the psychopathology of the average. If we don't have to change, we don't. We need merely review the track record of our New Year's resolutions. Unfortunately, in helping situations clients probably see most change as discretionary. They may talk about it as if were nondiscretionary, but deep down, the thought that "I don't really have to change" pervades the helping process. The sad state of discretionary change is not meant to discourage you but make you more realistic about the challenges you face — or the challenges you help your clients face.

Pervasiveness of Inertia

What keeps clients from acting on their own behalf? In physics, the law of inertia states, in part, that a body at rest tends to stay at rest unless something happens to move it along. Many clients are in trouble or stay in trouble because, for whatever reason, they are at rest. The sources of inertia are many, ranging from pure sloth to paralyzing fear. In trying to help clients act, you will come up against the many faces of inertia. We don't have to look far: most of us can start by examining inertia in our own lives.

We should not be too quick to blame our clients for their inertia. Inertia permeates life and is one of the principal mechanisms for keeping individuals, organizations, and institutions mired in the psychopathology of the average. A friend of mine is a consultant to organizations. He told me about one larger organization in which he worked with senior managers to transform the place. They did it right: diagnosis, identification of key issues, establishment of objectives, formulation of strategies, drawing up of plans. My friend went away satisfied with a job well done. He returned six months later to see how the changes were working out. But there were no changes. From the president on down, no one had acted on their plans. The managers looked sheepish and gave the excuses we all give — too busy, other things came up, couldn't get in touch with the other guy, ran into obstacles. And this was a fairly successful company.

The list of ways in which we avoid taking responsibility is endless. We'll examine several of them here: passivity, learned helplessness, disabling self-talk, getting trapped in vicious circles, and disorganization.

Passivity. One of the most important ingredients in the generation and perpetuation of the psychopathology of the average is passivity, the failure of people to take responsibility for themselves in one or more developmental areas of life or in various life situations that call for action. Passivity takes many different forms: (1) doing nothing — that is, not responding to problems and options; (2) uncritically accepting the goals and solutions suggested by others; (3) acting aimlessly; and (4) becoming incapacitated or violent — that is, shutting down or blowing up (*see* Schiff, 1975).

When Zelda and Jerzy first noticed small signs that things were not going right in their relationship, they did nothing. They noticed certain incidents, mused on them for a while, and then forgot about them. They lacked the communication skills to engage each other immediately and to explore what was happening. Zelda and Jerzy had both learned to remain passive before the little crises of life, not realizing how much their passivity would ultimately contribute to their downfall. Endless unmanaged problems led to major blow-ups until they decided to end their marriage.

Learned helplessness.   Seligman's (1975) concept of learned helplessness and its relationship to depression has received a great deal of attention since he first introduced it. Some clients learn to believe from an early age that there is nothing they can do about certain life situations. There are degrees in feelings of helplessness, from mild forms of "I'm not up to this" to feelings of total helplessness coupled with deep depression. Learned helplessness, then, is a step beyond mere passivity.

Bennett and Bennett (1984) saw the positive side of helplessness. If the problems clients face are indeed out of their control, then it is not helpful for them to have an illusory sense of control, unjustly assign themselves responsibility, and indulge in excessive expectations. Somewhat paradoxically, they found that challenging clients' tendency to blame themselves for everything actually fostered realistic hope and change. The trick is helping clients learn what is and what is not in their control. A man with a physical disability may not be able to do anything about the disability itself, but he does have some control over how he views his disability and his power to pursue certain life goals despite it.

The opposite of helplessness is resourcefulness. If helplessness can be learned, so can resourcefulness. Indeed, increased resourcefulness is one of the principal goals of successful helping.

Disabling Self-Talk.   Clients often get into the habit of engaging in disabling self-talk, thus talking themselves into passivity. They may say to themselves such things as "I can't cope, I can't do it, It won't work," or even "I don't have what it takes to engage in that program—it's too hard." Such self-defeating conversations get people into trouble in the first place and then prevent them from getting out. Ways of helping clients to manage their disabling self-talk will be discussed in Chapters 8 and 9.

Vicious circles.   Pyszczynski and Greenberg (1987) developed a theory about self-defeating behavior and depression. They said that people whose actions fail to get them what they want can easily lose a sense of self-worth and become mired in a vicious circle of guilt and depression.

> Consequently, the individual falls into a pattern of virtually constant self-focus, resulting in intensified negative affect, self-derogation, further negative outcomes, and a depressive self-focusing style. Eventually, these factors lead to a negative self-image, which may take on value by providing an explanation for the individual's plight and by helping the individual avoid further disappointments. The depressive self-focusing style then maintains and exacerbates the depressive disorder. (p. 122)

It does sound depressing. One client, Amanda, fits this theory perfectly. She had aspirations of moving up the career ladder where she worked. She was very enthusiastic and dedicated, but she was unaware of the "gentleman's club" politics of the company in which she worked and she didn't know how to work the system. She kept doing the things that she thought should get her ahead. They didn't. Finally, she got down on herself, began making mistakes in the things that she usually did well, and made things worse by constantly talking about how she "was stuck," thus alienating her friends. By the time she saw a counselor, she felt defeated and depressed. She was about to give up. The counselor focused on the entire circle — low self-esteem producing passivity producing even lower self-esteem — rather than on only the self-esteem part. Instead of just trying to help her change her inner world of disabling self-talk, he helped her intervene in her life to become a better problem solver. Small successes in problem solving led to the start of a benign circle — success producing greater self-esteem leading to greater efforts to succeed.

Disorganization.   I once knew someone who lived out of his car. No one knew exactly where he spent the night. The car was chaos, and so was his life. He was always going to get his career, family relations, and love life in order, but he never did. Living in disorganization was his way of putting off life decisions. Tom Ferguson (1987) painted a picture that may well remind us of ourselves, at least at times.

> When we saddle ourselves with innumerable little hassles and problems, they distract us from considering the possibility that we may have chosen the wrong job, the wrong profession, or the wrong mate. If we are drowning in unfinished housework, it becomes much easier to ignore the fact that we have become estranged from family life. Putting off an important project — painting a picture, writing a book, drawing up a business plan — is a way of protecting ourselves from the possibility that the result may not be quite as successful as we had hoped. Setting up our lives to insure a significant level of disorganization allows us to continue to think of ourselves as inadequate or partially-adequate people who don't have to take on the real challenges of adult behavior. (p. 46)

Many factors, such as defending ourselves against a fear of succeeding, can be behind this unwillingness to get our lives in order.

Driscoll (1984, pp. 112–117) has provided us with a great deal of insight into this problem. He described inertia as a form of control. He said that if we tell some clients to jump into the driver's seat, they will compliantly do so — at least until the journey gets too rough. The most effective strategy, he claimed, is to show clients that they have been in the driver's seat right along: "Our task as therapists is not to talk our clients into taking control of their lives, but to confirm the fact that they already are and always will be." The client who stays disorganized, sometimes even against great odds, is thus exercising a form of control and preserving inertia.

DEALING WITH INERTIA —
SELF-EFFICACY

Given the many faces of inertia, we cannot assume that clients will take action. In collaboration with them, we need to build action into the helping process right from the start. Understanding the processes of self-efficacy and self-regulation will help us do so. The opposite of passivity is agency, assertion or assertiveness (Galassi & Bruch, 1992), or self-efficacy (Bandura, 1977, 1980, 1982, 1986, 1989, 1991; Locke & Latham, 1990). A great deal of research continues to be done around the concept of self-efficacy and its applications to various settings, including education (*see* Multon, Brown, & Lent, 1991) and health care (O'Leary, 1992).

The Nature of Self-Efficacy

Bandura has suggested that people's expectations of themselves have a great deal to do with their willingness to put forth effort to cope with difficulties, the amount of effort they will expend, and their persistence in the face of obstacles. In particular, people tend to take action if two conditions are fulfilled:

1. They see that certain behavior will most likely lead to certain desirable results or accomplishments (outcome expectations).
2. They are reasonably sure that they can successfully engage in such behavior (self-efficacy expectations).

For instance, Yolanda believes that participation in a rather painful and demanding physical rehabilitation program following an accident and surgery will literally help her get on her feet again (an outcome expectation); she also believes that she has what it takes to inch her way through the program (a self-efficacy expectation). She therefore enters the program with a very positive attitude. Yves, on the other hand, is not convinced that an uncomfortable chemotherapy program will prevent his cancer from spreading and give him some quality living time (a negative outcome expectation), even though he knows he could endure it, so he says no to the doctors. Xavier is convinced that a series of radiation and chemotherapy treatments would help him (a positive outcome expectation), but he does not feel that he has the courage to go through with them (a negative self-efficacy expectation). He, too, refuses the treatment.

Helping Clients Develop Self-Efficacy

People's sense of self-efficacy can be strengthened in a variety of ways (*see* Mager, 1992). Lest self-efficacy be seen as a paradigm that applies only to the weak, let's consider some of these ways, used in the case of Nick, a very strong manager. Nick wanted to change his abrasive supervisory style, but was doubtful that he could do so; as he would say, "after all these years, I am what I am." It would have been silly just to say to him, "Nick, you can do it, believe in yourself." The fact that it was necessary to help him do a number of things to help strengthen his sense of self-efficacy does not mean that he was weak.

- *Skills*. Make sure that clients have the skills they need to perform desired tasks. Self-efficacy is based on ability. Nick first read about and then attended some skill-building sessions on such "soft" skills as listening, responding with empathy, giving feedback that was softer on the person and harder on the problem, and constructive challenging. In truth, he had many of these skills but they lay dormant. These short training experiences put him back in touch with some things he could do but didn't do.
- *Feedback*. Provide feedback that is based on deficiencies in performance, not on the deficiencies of the client. Since I attended many meetings with Nick, I routinely described his performance. This feedback was not about his personality but about his behavior.
- *Success*. Help clients not only act but see that their behavior actually produces results. Often success in a small endeavor will give them the courage to try something more difficult. Some of Nick's peers and direct reports began to notice changes in his behavior. He was pleased by this, but mentioned this only to me.
- *Modeling*. Help clients see others doing what they are trying to do and then encourage them to try themselves. I pointed out a few people in the company who were good at these "soft" skills but who were still considered competent, even tough, managers. On occasion, Nick would bring up instances of the soft approach working better than a hard one.
- *Encouragement*. Exhort clients to try, challenge them, and support their efforts. This should never be patronizing. Had I patronized Nick, I would have been dead; my encouragement of him was subtle and indirect.
- *Reducing fear and anxiety*. Help clients overcome their fears. If people are overly fearful that they will fail, they generally do not act. Therefore, procedures that reduce fear and anxiety help heighten the sense of self-efficacy. Deep down Nick was fearful of two things regarding changing his supervisory style — being less effective in managing the business and making a fool of himself. Business results helped allay the former. He even noticed that two of his direct reports seemed to become more productive. As to the latter, his behavior outside the office came to the rescue. When we visited plants and field offices, Nick was very upbeat. He was as good at rallying the troops as anyone I have ever seen. Discussions about his two different styles helped allay fears that he would make a fool of himself by changing his style.

As a helper, you can do a great deal to help people develop a sense of agency or self-efficacy. First of all, you can help them challenge self-defeating beliefs and attitudes about themselves and substitute realistic beliefs about their ability to act. This includes helping them reduce the kinds of fears and anxieties that keep them from mobilizing their resources. Second, you can help them develop the working knowledge, life skills, and resources they need to succeed. Third, you can help them challenge themselves to take reasonable risks and support them when they do.

THE SELF-HELP MOVEMENT

We live in a society that has become somewhat suspicious of and confrontational toward professional help. Further, there is a growing realization that help must begin at home; that is, with the person with the problem. Thus the self-help movement (Gartner & Riessman, 1977, 1984; Pancoast, Parker, & Froland, 1983; Riessman, 1985).

Self-Help Groups

The self-help movement is flourishing (Harvard, 1993a, 1993b; Katz & Bender, 1990). Across the country groups have been established to help people cope with almost every conceivable problem. In 1992 a study conducted by Princeton's Center for the Study of American Religion found that there are some 3 million small groups meeting regularly in the United States (*see* Lattin, 1992). Some of the groups focus on spiritual self-help and constitute what the Center's Robert Wuthnow has called a kind of do-it-yourself religion. Others provide support for all the problems that beset humankind. Admittedly, it is difficult to see where spiritual support ends and psychological support begins.

Some twenty-five years ago, Hurvitz (1970, 1974) claimed that these groups were more effective and efficient than other forms of helping: "It is likely that more people have been and are being helped by [self-help groups] than have been and are being helped by all types of professionally trained psychotherapists combined, with far less theorizing and analyzing and for much less money" (1970, p. 48). Wuthnow's study certainly provides support for this hypothesis today. The size and robustness of this movement—the average group lasts some five years—might surprise a lot of professionals because most groups are quiet, small, and local. They range from such traditional groups as Alcoholics Anonymous to groups that help women cope with the aftermath of a mastectomy operation, from Weight Watchers to newly formed groups of people who are HIV-positive but have not yet developed AIDS.

The Self-Help Bias toward Action

Research suggests that self-help groups differ greatly in their approaches to helping (McFadden, Seidman, & Rappaport, 1992). Yet the pragmatic, action-oriented, democratic ethos of self-help groups, as described in an issue of the *Self-Help Reporter* (Fall, 1985, p. 5), are congruent with the value approach taken in this book, including:

- A noncompetitive, cooperative orientation
- An anti-elite, anti-bureaucratic focus
- An experiential emphasis — people who have the problem know a lot about it from the inside, from experiencing it
- A pragmatic do-what-you-can-do, one-day-at-a-time, and you-can't-solve-everything-at-once approach
- A shared, often rotating leadership
- Being helped by helping others
- A refusal to see helping as a commodity to be bought and sold
- An accent on using one's own power in taking control over one's own life
- A strong optimism regarding the ability to change
- A belief that small steps are important
- A critical stance toward professionalism, which is often pretentious, purist, distant, and mystifying
- Simplicity and informality
- The centrality of helping — knowing how to receive help, give help, and help yourself
- An avoidance of self-victimization
- A focus on the group with getting back into community as essential

Of course, any given individual or self-help group can deviate from the self-responsibility and action focus just outlined. Some people become obsessed with their own disabilities and the self-help group itself — another form of addiction — and this undermines the capacity for personal action (*see* Kaminer, 1992). On the other hand, there are many examples of individuals and groups that have used the self-help forum as a basis for social and political action (see *Self Help Reporter*, Summer, 1992, p. 1). In general, those who join self-help groups are motivated to take active roles in changing their lives by the group's peer pressure. Even passive group members can be moved to action. In the past, professional helpers tended to ignore self-help groups, perhaps as a reaction to the movement's suspicion of professionals. Now, more and more professionals are taking self-help groups seriously. Given the inexhaustibility of human need, there is a place for professionals, for the self-help movement, and for partnerships between the two.

HELPERS AS AGENTS

Driscoll (1984, pp. 91–97) discussed the temptation of helpers to respond to the passivity of their clients with a kind of passivity of their own, a "Sorry, it's up to you" stance. This, he claimed, is a mistake.

> A client who refuses to accept responsibility thereby invites the therapist to take over. In remaining passive, the therapist foils the invitation, thus forcing the client to take some initiative or to endure the silence. A passive stance is therefore a means to avoid accepting the wrong sorts of responsibility. It is

generally ineffective, however, as a long-run approach. Passivity by a therapist leaves the client feeling unsupported and thus further impairs the already fragile therapeutic alliance. Troubled clients, furthermore, are not merely unwilling but generally and in important ways unable to take appropriate responsibility. A passive countermove is therefore counterproductive, for neither therapist nor client generates solutions, and both are stranded together in a muddle of entangling inactivity. (p. 91)

In order to help others act, helpers must be agents and doers in the helping process, not mere listeners and responders. The best helpers are active in the helping sessions. They keep looking for ways to enter the worlds of their clients, to get them to become more active in the sessions, to get them to own more of it, to help them see the need for action in their heads and outside their heads in their everyday lives. And they do all this without violating the values outlined in Chapter 3. They don't push reluctant clients, thus turning reluctance into resistance; neither do they sit around waiting for reluctant clients to act. The rest of this book, then, is action-oriented, oriented to action that is both purposeful and prudent.

PART TWO

BASIC COMMUNICATION SKILLS FOR HELPING

Since helping takes place through a dialogue between helper and client, the communication skills of the helper are critical tools. The basic communication skills—attending, listening, empathy, and probing—are outlined, discussed, and illustrated in Chapters 5 and 6.

CHAPTER 5

COMMUNICATION SKILLS I: ATTENDING AND LISTENING

The Importance of
Communications Skills

Since the helping process involves a great deal of communication between helper and client, relevant communication skills are extremely important for the helper at every stage and step of the helping process. These skills are not the helping process itself, but they are essential tools for developing relationships and interacting with clients. This chapter deals with the first set of these skills, attending and listening. Chapter 6 deals with a second set, empathy and probing. Chapter 9 deals with a third set related to helping clients identify blind spots and develop new perspectives.

These skills are not special skills peculiar to helping. Rather, they are extensions of the kinds of skills all of us need in our everyday interpersonal transactions. Ideally, helpers-to-be would enter training programs with this basic set of interpersonal communication skills in place, and training would simply help them adapt the skills to the helping process. Unfortunately, this is often not the case. That is why these skills are reviewed here before the helping process itself is explored in greater detail.

Since communication skills are not ends in themselves but means or instruments to be used in achieving helping outcomes, there has been a growing concern about the overemphasis of communication skills and techniques. Years ago Rogers (1980) spoke out against what he called the "appalling consequences" (p. 139) of an overemphasis on the microskills of helping. Instead of being a fully human endeavor, helping was, in his view, being reduced to its bits and pieces. Some helper training programs focus almost exclusively on these skills. As a result, trainees know how to communicate but not how to help.

Hills (1984) discussed an integrative versus a technique approach to training in communication skills. In an integrative approach, skills and techniques become extensions of the helper's humanity and not just bits of helping technology; further, these communication skills and helping techniques serve the goals of the helping process. The skills and techniques of an integrative approach are permeated with and driven by the values discussed in Chapter 3. In short, skilled trainees will learn these communication skills, practice them, and use them in a fully human way in supervised sessions with clients.

In helping interviews, the skills reviewed in this chapter and in Chapter 6 are woven together at the service of the client. However, for the purpose of training, they are treated separately here. To begin, we need to distinguish between attending and listening:

- Attending refers to the ways in which helpers can be with their clients, both physically and psychologically.
- Listening refers to the ability of helpers to capture and understand the messages clients communicate, whether these messages are transmitted verbally or nonverbally, clearly or vaguely.

These skills, including some of the philosophy underlying them, are now examined in some detail.

ATTENDING: ACTIVELY BEING WITH CLIENTS

At some of the more dramatic moments of life, simply being with another person is extremely important. If a friend of yours is in the hospital, just being there can make a difference, even if conversation is impossible. Similarly, just being with a bereaved friend can be very comforting to him or her, even if little is said. People appreciate it when others pay attention to them. By the same token, being ignored is often painful: the averted face is too often a sign of the averted heart. Given how sensitive most of us are to others' attention or inattention, it is paradoxical how insensitive we can be at times about attending to others.

Helping and other deep interpersonal transactions demand a certain intensity of presence. Attending, or the way you orient yourself physically and psychologically to clients, contributes to this presence. Effective attending does two things: it tells clients that you are with them, and it puts you in a position to listen carefully to their concerns.

Clients read cues that indicate the quality of your presence to them. Your nonverbal behavior influences clients for better or worse. Attentive presence can invite or encourage them to trust you, open up, and explore the significant dimensions of their problem situations. Halfhearted presence can promote distrust and lead to clients' reluctance to reveal themselves to you.

There are various levels of attending to clients: (1) the microskills level, (2) the body language level, and (3) the human presence level.

The Microskills of Attending

There are certain microskills that helpers can use in attending to clients. Although these skills represent only the first level of attending or being with clients, they do serve as a starting point. The microskills can be summarized in the acronym *SOLER*.

- *S*: Face the client *Squarely*; that is, adopt a posture that indicates involvement. In North-American culture, facing another person squarely is often considered a basic posture of involvement. It usually says: "I'm here with you; I'm available to you." Turning your body away from another person while you talk to him or her can lessen your degree of contact with that person. Even when people are seated in a circle, they usually try in some way to turn toward the individuals to whom they are speaking. The word *squarely* here may be taken literally or metaphorically. What is important is that the bodily orientation you adopt convey the message that you are involved with the client. If, for any reason, facing the person squarely is too threatening, then an angled position may be more helpful. The point is the quality of your attention.

- *O*: Adopt an *Open* posture. Crossed arms and crossed legs can be signs of lessened involvement with or availability to others. An open posture can be a sign that you're open to the client and to what he or she has to say. In North-American culture, an open posture is generally seen as a nondefensive posture. Again, the word *open* can be taken literally or metaphorically. If your legs are crossed, this does not mean that you are not involved with the client. But it is important to ask yourself: "To what degree does my present posture communicate openness and availability to the client?"

- *L*: Remember that it is possible at times to *Lean* toward the other. Watch two people in a restaurant who are intimately engaged in conversation. Very often they are both leaning forward over the table as a natural sign of their involvement. The main thing is to remember that the upper part of your body is on a hinge. It can move toward a person and back away. In North-American culture, a slight inclination toward a person is often seen as saying, "I'm with you, I'm interested in you and in what you have to say." Leaning back (the severest form of which is a slouch) can be a way of saying, "I'm not entirely with you" or "I'm bored." Leaning too far forward, however, or doing so too soon, may frighten a client. It can be seen as a way of placing a demand on the other for some kind of closeness or intimacy. In a wider sense, the word *lean* can refer to a kind of bodily flexibility or responsiveness that enhances your communication with a client.

- *E*: Maintain good *Eye* contact. In North-American culture, fairly steady eye contact is not unnatural for people deep in conversation. It is not the same as staring. Again, watch two people deep in conversation. You may be amazed at the amount of direct eye contact. Maintaining good eye contact with a client is another way of saying, "I'm with you; I want to hear what you have to say." Obviously this principle is not violated if you occasionally look away. Indeed you have to if you don't want to stare. But if you catch yourself looking away frequently, your behavior may give you a hint about some kind of reluctance to be with this person or to get involved with him or her. Or it may say something about your own discomfort.

- *R*: Try to be relatively *Relaxed* or natural in these behaviors. Being relaxed means two things. First, it means not fidgeting nervously or engaging in distracting facial expressions. The client may wonder what's making you nervous. Second, it means becoming comfortable with using your body as a vehicle of contact and expression. Being natural in the use of skills helps put the client at ease.

These guidelines should be adopted cautiously, especially in multicultural counseling settings (Sue, 1990). People differ both individually and culturally in how they show attentiveness. The main point is that an internal mind-set in which you are with a client might well lose its impact if the client does not see this internal attitude reflected in your nonverbal communication. It is not uncommon for helpers in training to become overly self-conscious about their attending behavior, especially in the beginning and perhaps even more especially if they are not used to attending carefully to others. The guidelines presented are just that—guidelines—don't take them as absolute rules to be applied rigidly in all cases.

Nonverbal Communication

Much more important than a mechanical application of microskills is an awareness of your body as a source of communication. Effective helpers are mindful of the cues and messages they are constantly sending through their bodies as they interact with clients. Reading your own bodily reactions is an important first step. For instance, if you feel your muscles tensing as the client talks to you, you can say to yourself: "I'm getting anxious here. What's causing my anxiety? And what cues am I sending the client?" Once you read your own reactions, you can use your body to communicate appropriate messages. You can also use your body to censor messages that you feel are inappropriate. For instance, if the client says something that instinctively angers you, you can control the external expression of the anger (for instance, a sour look) to give yourself time to reflect. This second level of attending does not mean that you become preoccupied with your body as a source of communication. Rather, it means that you learn to use your body as a means of communication. Being aware of and at home with nonverbal communication can reflect an inner peace with yourself, the helping process, and this client.

Social-Emotional Presence

Most important is the quality of your total human presence to your clients. Both your verbal and your nonverbal behavior should indicate a clear-cut willingness to work with the client. If you care about your clients and feel committed to their welfare, then it is unfair to yourself to let your nonverbal behavior suggest contradictory messages. On the other hand, if you feel indifference about them and your nonverbal behavior suggests commitment, then you are not being genuine. Effective helpers stay in touch with how they present to clients without becoming preoccupied with it.

Box 5-1 summarizes, in question form, the main points related to attending, especially in terms of social-emotional presence. Obviously, helpers are not constantly asking these questions of themselves as they interact with clients, but they are in touch with the quality of their presence to their clients.

BOX 5-1

Questions on Attending

- What are my attitudes toward this client?
- How would I rate the quality of my presence to this client?
- To what degree does my nonverbal behavior indicate a willingness to work with the client?
- What attitudes am I expressing in my nonverbal behavior?
- What attitudes am I expressing in my verbal behavior?
- To what degree does the client experience me as effectively present and working with him or her?
- To what degree does my nonverbal behavior reinforce my internal attitudes?
- In what ways am I distracted from giving my full attention to this client?
- What am I doing to handle these distractions?
- How might I be more effectively present to this person?

Active Listening

Effective attending puts helpers in a position to listen carefully to what clients are saying both verbally and nonverbally. Listening carefully to a client's concerns seems to be a concept so simple to grasp and so easy to do that one may wonder why it is given such explicit treatment here. Nonetheless, it is amazing how often people fail to listen to one another. How many times have you heard someone exclaim, "You're not listening to what I'm saying!" When the person accused of not listening answers, almost predictably, "I am, too; I can repeat everything you've said," the accuser is not comforted. What people look for in attending and listening is not the other person's ability to repeat their words. A tape recorder could do that perfectly. People want more than physical presence in human communication; they want the other person to be present psychologically, socially, and emotionally.

Complete listening involves four things: first, observing and reading the client's *nonverbal* behavior — posture, facial expressions, movement, tone of voice, and the like. Second, listening to and understanding the client's *verbal* messages. Third, listening to the *context;* that is to the whole person in the context of the social settings of his or her life. Fourth, listening to *sour notes;* that is, things the client says that may have to be challenged.

Listening to and Understanding
Nonverbal Behavior

Clients send messages through their nonverbal behavior. The skill helpers need is to read these messages without distorting or overinterpreting them.

Nonverbal behavior as a channel of communication.   Over the years re-searchers and practitioners have come to appreciate the importance of non-verbal behavior in counseling (Siegman & Feldstein, 1987). This area has taken on even more importance because of the multicultural nature of help-ing (Sue, 1990). The face and body are extremely communicative. We know from experience that, even when people are together in silence, the atmo-sphere can be filled with messages. Sometimes the facial expressions, bodily motions, voice quality, and physiological responses of clients com-municate more than their words. An early study illustrates this point. Mehrabian (1971) wanted to know what cues people use to judge whether another person likes them or not. He and his associates discovered that the other person's actual words contributed only 7% to the impression of being liked or disliked; voice cues contributed 38%; and facial cues, 55%. They also discovered that when facial expressions were inconsistent with spoken words, facial expressions were believed more than the words.

What is significant in Mehrabian's research is not the exact percentages but rather the clear importance of nonverbal behavior in the communication process. Effective helpers learn how to listen to and read the following:

- *bodily behavior*, such as posture, body movements, and gestures
- *facial expressions*, such as smiles, frowns, raised eyebrows, and twisted lips
- *voice-related behavior*, such as tone of voice, pitch, voice level, intensity, inflection, spacing of words, emphases, pauses, silences, and fluency
- *observable autonomic physiological responses*, such as quickened breathing, the development of a temporary rash, blushing, paleness, and pupil dilation
- *physical characteristics*, such as fitness, height, weight, complexion, and the like
- *general appearance*, such as grooming and dress

Once more, the trick is to spot the messages in these behaviors without making too little or too much of them.

The following case will be used to help you get a better feeling for both attending and listening.

Jennie, an African-American college senior, was raped by a supposed friend, while on a date. She received some immediate counseling from the university Student Development Center and some ongoing support during the subse-quent investigation. She knew she had been raped, but it was impossible for her to prove her case. The entire experience — both the rape and the investiga-tion that followed — left her shaken, unsure of herself, angry, and mistrustful of institutions she had assumed would be on her side (especially the university and the legal system). When Denise, a counselor for a health maintenance organization (HMO) first saw her a few years after the incident, Jennie was plagued by a number of somatic complaints, including headaches and gastric problems. At work she engaged in angry outbursts whenever she felt that someone was taking advantage of her. Otherwise she had become quite passive and chronically depressed. She saw herself as a woman victimized by society and was slowly giving up on herself.

When Denise said to Jennie, "It's hard talking about yourself, isn't it?" Jennie said, "No, I don't mind at all." But the real answer was probably in her nonverbal behavior, for she spoke hesitatingly while looking away and frowning. Reading such cues helped Denise understand Jennie better. A person's nonverbal behavior has a way of leaking messages to others. The very spontaneity of nonverbal behaviors contributes to this leakage even in the case of highly defensive clients. It is not easy for clients to fake nonverbal behavior (Wahlsten, 1991). The real messages still tend to leak out.

Nonverbal behavior as punctuation.   Besides being a channel of communication in itself, such nonverbal behavior as facial expressions, bodily motions, and voice quality often punctuate verbal messages in much the same way that periods, question marks, exclamation points, and underlining punctuate written language. Nonverbal behavior can punctuate or modify interpersonal communication in the following ways (*see* Knapp, 1978, pp. 9–12):

- *Confirming or repeating*. Nonverbal behavior can confirm or repeat what is being said verbally. For instance, once when Denise responded to Jennie with just the right degree of understanding, Jennie not only said, "That's right!" but her eyes lit up (facial expression), she leaned forward a bit (bodily motion), and her voice was very animated (voice quality). Her nonverbal behavior confirmed her verbal message.
- *Denying or confusing*. Nonverbal behavior can deny or confuse what is being said verbally. When challenged by Denise once, Jennie denied that she was upset, but her voice faltered a bit (voice quality) and her upper lip quivered (facial expression). Her nonverbal behavior carried the real message.
- *Strengthening or emphasizing*. Nonverbal behavior can strengthen or emphasize what is being said. When Denise suggested to Jennie that she might discuss the origin of what her boss saw as erratic behavior, Jennie said, "Oh, I don't think I could do that!" while slouching down and putting her face in her hands. Her nonverbal behavior underscored her verbal message. Nonverbal behavior adds emotional color or intensity to verbal messages. Once Jennie told Denise that she didn't like to be confronted without first being understood and then stared at her fixedly and silently with a frown on her face. Jennie's nonverbal behavior told Denise something about the intensity of her feelings.
- *Controlling or regulating*. Nonverbal cues are often used in conversation to regulate or control what is happening. If, in group counseling, one participant looks at another and gives every indication that she is going to speak to this other person, she may hesitate or change her mind if the per-

son to whom she intends to talk looks away. Skilled helpers are aware of the ways in which clients send controlling or regulating nonverbal cues.

Of course, helpers can punctuate their words with the same kind of nonverbal messages. Denise, too, without knowing it, sent silent messages to Jennie. For instance, she was especially attentive when Jennie talked about actions she could take to do something about her problem situation. This was part of the social-influence dimension of helping.

In reading nonverbal behavior — note that the word *reading* is used here instead of *interpreting* — caution is a must. We listen in order to understand clients rather than to dissect them. There is no simple program available for learning how to read and then interpret nonverbal behavior. Once you develop a working knowledge of nonverbal behavior and its possible meanings, you must learn through practice and experience to be sensitive to it and read its meaning in any given situation.

Since nonverbal behaviors can often mean a number of things, how can you tell which meaning is the real one? The key is the human context in which they take place. Effective helpers listen to the entire context of the helping interview and do not become overly fixated on details of behavior. They are aware of and use the nonverbal communication system, but they are not seduced or overwhelmed by it. This is the integrative approach. Sometimes novice helpers will fasten selectively on this or that bit of nonverbal behavior. For example, they will make too much of a client's half-smile or frown. They will seize upon the smile or the frown and, in overinterpreting it, lose the person.

Listening to and Understanding Verbal Messages

Beyond nonverbal cues and messages, what do helpers listen to? As outlined in Chapter 4, they listen to clients' verbal descriptions of their *experiences, behaviors,* and *affect.* A problem situation is clear if it is understood in terms of specific experiences, specific behaviors, and specific feelings and emotions. The counselor's job is to help clients achieve this kind of clarity. For example, to return to our case, Denise listened to what Jennie had to say early on about her past and present experiences, actions, and emotions. Jennie told her: "I had every intention of pushing my case, because I knew that men on campus were getting away with murder. But then it began to dawn on me that people were not taking me seriously because I was an African-American woman. First I was angry, but then I just got numb. . . ." Later, Jennie said, "I get headaches a lot now. I don't like taking pills, so I try to tough it out. I have also become very sensitive to any kind of injustice, even in movies or on television. But I've stopped being any kind of crusader. That got me nowhere."

Denise heard Jennie's disillusionment and wondered whether Jennie was disillusioned not only with society and its institutions but also with herself. She listened very carefully, because she realized that people had not listened to and believed Jennie when she was telling the truth. She listened for verbal and nonverbal messages. She listened to the feelings and emotions that permeated Jennie's words and nonverbal behavior. As she listened to Jennie speak, she constantly asked herself such questions as: "What are the core messages here? What themes are coming through? What is Jennie's point of view? What is most important to her? What does she want me to understand?" That is, Denise listened actively. Her first instinct was not to formulate responses as Jennie spoke, but just to listen.

Listening to and Understanding Clients in Context

People are more than the sum of their verbal and nonverbal messages. Listening in its deepest sense means listening to clients themselves as influenced by the contexts in which they live, move, and have their being. Denise tried to understand Jennie's verbal and nonverbal messages, even the core messages, in terms of the context of Jennie's life. As she listened to Jennie's story, Denise said to herself:

> Here is an intelligent African-American woman from a conservative Catholic background. She is very loyal to the church because it proved to be an inner-city refuge. It was a gathering place for her family and friends. It meant a decent primary- and secondary-school education and a shot at college. Initially college was a shock. It was her first venture into a predominantly white and secular culture. But she chose her friends carefully and carved out a niche for herself. Studies were much more demanding, and she had to come to grips with the fact that, in this larger environment, she was closer to average. The rape and investigation put a great deal of stress on what proved to be a rather fragile social network. Her life began to unravel. She pulled away from her family, her church, and the small circle of friends she had at college. At a time she needed support the most, she cut it off. After graduation, she continued to stay "out of community." She got a job as a secretary in a small company and has remained underemployed.

In other words, Denise tried to pull together the themes she saw emerging in Jennie's story and to see these themes in context.

Denise strove to listen to Jennie in the context of her social background. She listened to Jennie's discussion of her headaches (experiences), her self-imposed social isolation (behaviors), and her chronic depression (affect) against the background of her social history — the pressures of being religious in a secular society, the problems associated with being an upwardly mobile African-American woman in a society dominated by white males. Denise sees the rape and investigation as social, not merely personal, events. She listens actively and carefully, because she knows that her ability to help depends, in part, on not distorting what she hears. She does not focus narrowly on Jennie's inner psychology, as if Jennie could be separated from the social context of her life.

It is also important to listen to clients in terms of the helping process itself. Not all clients start in the first step of the first stage. Therefore, you need to know where the client is in the model itself if you are to be fully helpful. For instance, let's say that a client says something like this during the first interview:

> I could not have come here before now. You know why? Because for the first time I've said to myself that I am too promiscuous for my own good. For my partners' good. If there is such a thing as a sex addict, then I'm probably one. I've got to do something about this. I'm endangering myself and others.

Where is the client in the helping model? He is in some kind of combination of Stages II and III. He wants to develop a different approach to his sex life (Stage II) and he wants some help in doing whatever is necessary to develop this new approach (Stage III).

Ineffective helpers fail to listen to where their clients are. They listen to themselves, they listen to their models. They drag their clients through their models. For instance, if a helper were to reply to the client in this example by saying, "Well, now, let's explore the issues. Let's see what your concerns are," he would be responding as if the client were in Stage I. In contrast, skilled helpers respond to clients where they are in the helping process. Here a skilled helper might say, "It's really come home to you. It's time to change your behavior." Later, it may be necessary to double back to an earlier stage or step, but the client dictates the starting point. Clients may move from stage to stage in the helping process. Careful listening enables helpers to stay with their clients. Finally, since problem-managing action is essential, helpers need to listen carefully to clients' readiness for action.

Tough-Minded Listening

Clients' visions of and feelings about themselves, others, and the world are real and need to be understood. However, their perceptions of themselves and their worlds are sometimes distorted. For instance, if a client sees herself as ugly, when in reality she is beautiful, her experience of herself as ugly is real and needs to be listened to and understood. But her experience of herself does not square with the facts. This, too, must be listened to and understood. If a client sees himself as above average in his ability to communicate with others when, in reality, he is below average, his experience of himself needs to be listened to and understood, but reality cannot be ignored. Tough-minded listening includes detecting the gaps, distortions, and dissonance that are part of the client's experienced reality. This does not mean that helpers challenge clients as soon as they hear any kind of distortion. Rather, they note gaps and distortions and challenge them when it is appropriate to do so (see Chapters 8 and 9).

Denise realized from the beginning that some of Jennie's understandings of herself and her world were not accurate. For instance, in reflecting on all that happened, Jennie remarked that she probably got what she deserved. When Denise asked what she meant, Jennie said, "My ambitions were too high. I was getting beyond my place in life." It is one thing to understand how Jennie might put this interpretation on what has happened; it is another to assume that such an interpretation reflects reality. To be client-centered, helpers must first be reality-centered.

THE SHADOW SIDE OF LISTENING TO CLIENTS

Active listening is not as easy as it sounds. Obstacles and distractions abound. As you will recognize from your own experience, the following kinds of ineffective listening overlap with one another.

Inadequate listening.   In conversations it is easy for us to be distracted from what other people are saying. We get involved in our own thoughts, or we begin to think about what we are going to say in reply. At such times we may then hear "You're not listening to me!" Helpers, too, can become preoccupied with themselves and their own needs in such a way that they are kept from listening fully to their clients. They are attracted to their clients, they are tired or sick, they are preoccupied with their own problems, they are too eager to help, they are distracted because clients have problems similar to their own, or the social and cultural differences between them and their clients make listening and understanding difficult. It is sometimes difficult to remember that each client is new and unique and that each session with the same client is a new challenge.

Evaluative listening.   Most people, even when they listen attentively, listen evaluatively. That is, as they listen, they are judging the merits of what the other person is saying in terms of good-bad, right-wrong, acceptable-unacceptable, like-dislike, relevant-irrelevant, and so forth. Helpers are not exempt from this universal tendency. The following interchange took place between Jennie and a friend of hers. Jennie recounted it to Denise as part of her story.

JENNIE: Well, the rape and the investigation are not dead, at least not in my mind. They are not as vivid as they used to be, but they are there.
FRIEND: That's the problem, isn't it? Why don't you do yourself a favor and forget about it? Get on with life, for God's sake!

This might well be sound advice, but the point here is that Jennie's friend listened and responded evaluatively. Clients should first be understood; then, if necessary, they should be challenged or helped to challenge themselves. Evaluative listening, translated into advice giving, will just put clients off. Of course, understanding the client's point of view is not the same as accepting it. Indeed, a judgment that a client's point of view, once understood, needs to be expanded or transcended or that a pattern of behavior, once listened to and understood, needs to be altered can be quite useful. That is, there are productive forms of evaluative listening. It is practically impossible to suspend judgment completely. Nevertheless, it is possible to set one's judgment aside for the time being at the service of understanding clients, their worlds, and their points of view.

Filtered listening.   It is impossible to listen to other people in a completely unbiased way. Through socialization we develop a variety of filters through which we listen to ourselves, others, and the world around us. As Hall (1977) noted: "One of the functions of culture is to provide a highly selective screen between man and the outside world. In its many forms, culture therefore designates what we pay attention to and what we ignore. This screening provides structure for the world" (p. 85). We need filters to provide structure for ourselves as we interact with the world. But personal, familial, sociological, and cultural filters introduce various forms of bias into our listening and do so without our being aware of it.

The stronger the cultural filters, the greater the likelihood of bias. For instance, a white, middle-class helper probably tends to use white, middle-class filters in listening to others. Perhaps this makes little difference if the client is also white and middle class. But if the client is a well-to-do Oriental with high social status in his community, an African-American mother from an urban ghetto, or a poor white subsistence farmer, then the helper's cultural filters might introduce bias. Prejudices, whether conscious or not, distort understanding. Like everyone else, helpers are tempted to pigeonhole clients based on such factors as gender, race, sexual orientation, nationality, social status, religious persuasion, political preferences, and lifestyle. In Chapter 1 the importance of self-knowledge on the part of the helper was noted. This includes ferreting out the biases and prejudices that distort our listening.

Labels as filters.   Sometimes book learning can act as a distorting filter. Books on personality theories provide us with pigeonholes. We even pigeonhole ourselves. Diagnostic categories such as schizophrenia can take precedence over the persons being diagnosed. If what you hear is the theory and not the person, you can be technically correct in your diagnosis and once more lose the client. In short, what you learn as you study psychology may help you to organize what you hear — but it may also distort your listening. To use terms borrowed from Gestalt psychology, make sure that your client remains figure (in the forefront of your attention) and that models and theories about clients remain ground (learnings that remain in the background and are used only in the service of understanding and helping this unique client).

Fact-centered rather than person-centered listening.   Some helpers ask a lot of informational questions, as if the client would be cured if enough facts about him or her were known. It's entirely possible to collect facts but miss the person. The antidote is to listen to clients contextually, trying to focus on themes and key messages. In listening to Jennie, Denise picked up what is called the "pessimistic explanatory style" theme (Peterson, Seligman, & Vaillant, 1988). This is the tendency to attribute causes to negative events that are stable ("It will never go away"), global ("It affects everything I do"), and internal ("It is my fault"). Research indicates that people who fall victim to this style tend to end up with poorer health than those who do not. Denise knew of this research and therefore hypothesized that there might be a link between Jennie's somatic complaints (headaches, gastric problems) and this explanatory style. This is a theme worth exploring.

Rehearsing.   When beginning helpers ask themselves, "How am I to respond to what the client is saying?" they stop listening. When experienced helpers begin to mull over what might be the perfect response to what their clients are saying, they stop listening. Helping is more than the technology found in these pages. It is also an art. Helpers who listen intently to clients and to the themes and core messages embedded in what they are saying, however haltingly or fluently they say it, are never at a loss in responding. They don't need to rehearse. And their responses are much more likely to help clients move forward in the problem-management process. When the client stops speaking, they often pause to reflect on what he or she just said and then speak. But this is not rehearsing.

Sympathetic listening.   Often clients are people in pain or people who have been victimized by others or by society itself. Such clients can arouse feelings of sympathy in helpers. Sometimes these feelings are strong enough to distort the stories that are being told. Recall the case of Ben, the client who lost his wife and daughter, and his helper, Liz.

> Liz had recently lost her husband to cancer. During their first meeting, as Ben talked about his own tragedy, she wanted to hold him. Later that day she took a long walk and realized how her sympathy for Ben distorted what she heard. She heard the depth of his loss. But, reminded of her own loss, she only half heard the implication that this now excused him from many of the tasks of life.

Sympathy has an unmistakable place in human transactions, but its usefulness, if that does not sound too inhuman, is limited in helping. In a sense, when I sympathize with someone, I become his or her accomplice. If I sympathize with my client as she tells me how awful her husband is, I take sides without knowing what the complete story is. Helpers should not become accomplices in letting a client's self-pity drive out problem-managing action.

Interrupting. I am reluctant to add interrupting to this list of obstacles. Certainly, by interrupting clients, the helper stops listening. And interrupters often deliver messages they have been rehearsing. My reluctance, however, comes from the conviction that helping goes best when it is a dialogue between client and helper. I seldom find monologues, including my own, helpful. Occasionally, monologues that help clients get their stories or significant updates of their stories out are useful. Even then, such a monologue is best followed by a fair amount of dialogue. Therefore, I see benign and malignant forms of interrupting. The helper who cuts the client off in mid-thought because he has something important to say is using a malignant form. But the case is different when a helper interrupts a monologue with some gentle gesture and a comment such as, "You've made several points. I want to make sure that I've understood them." If interrupting promotes the kind of dialogue that serves the problem-management process, then it is useful.

LISTENING TO ONESELF

To be an effective helper, you need to listen not only to the client but to yourself. Granted, you don't want to become self-preoccupied, but listening to yourself — on what you might think of as a second channel — can help you identify what is standing in the way of your being with and listening to the client. It is a positive form of self-consciousness.

Some years ago this second channel did not work very well for me. A friend of mine who had been in and out of mental hospitals for a few years and whom I had not seen for over six months showed up one evening at my apartment unannounced. He was in a highly excited state. A torrent of ideas, some outlandish, some brilliant, flowed nonstop from him. I sincerely wanted to be with him as best I could, but I was very uncomfortable. I started by more or less naturally following the rules of attending, but I kept catching myself sitting with my arms and legs crossed. I think that I was defending myself from the torrent of ideas. When I discovered myself almost literally tied up in knots, I would untwist my arms and legs, only to find them crossed again a few minutes later. It was hard work being with him. In retrospect, I realize I was concerned for my own security. I have since learned to listen to myself on the second channel a little better — to my nonverbal behaviors as well as my internal dialogues — so that these interactions might serve clients better.

With a somewhat different emphasis, Rogers (1980) talked about letting oneself get lost in the world of the client.

> In some sense [attending and listening] means that you lay aside your self; this can only be done by persons who are secure enough in themselves that they know they will not get lost in what may turn out to be the strange and bizarre

world of the other, and that they can comfortably return to their own world when they wish. (p. 143)

This ability to focus almost exclusively on the client while forgetting oneself and then return to productive self-consciousness comes with both experience and maturity. Keeping an ear lightly tuned to oneself is certainly a step short of the kind of total immersion Rogers described, but, as we shall see in greater detail later, it can serve the client's goals.

Box 5-2 summarizes, in question form, the main points related to effective listening.

BOX 5-2

QUESTIONS ON LISTENING

- How well do I read the client's nonverbal behaviors and see how they modify what he or she is saying verbally?
- How careful am I not to overinterpret nonverbal behavior?
- How intently do I listen to what the client is saying verbally, noticing the mix of experiences, behaviors, and feelings?
- How effectively do I listen to the client's point of view, especially when I sense that this point of view needs to be challenged or transcended?
- How easily do I tune in to the core messages being conveyed by the client?
- How effective am I at spotting themes in the client's story?
- What distracts me from listening more carefully? What can I do to manage these distractions?
- How effectively do I pick up cues indicating dissonance between reality and what the client is saying?
- To what degree can I note the ways in which the client exaggerates, contradicts himself or herself, misinterprets reality, and holds things back without judging him or her and without interfering with the flow of the dialogue?
- How effectively do I listen to what is going on inside myself as I interact with clients?

COMMUNICATION SKILLS II: BASIC EMPATHY AND PROBING

THE ART OF EMPATHY

As we have seen, attending and listening are not passive activities. They are part of the helper's bias toward action. The fruit of attending and listening lies in the way the helper responds to the client. Therefore, it is essential to move on to the responding skills of empathy and probing. Empathy as a form of human communication involves listening to clients, understanding them and their concerns to the degree that this is possible, and communicating this understanding to them so that they might understand themselves more fully and act on their understanding. Empathy that remains locked up in the helper contributes little to the helping process.

There is still some confusion among helpers as to just what empathy is. Some of the confusion comes from the distinction between empathy as a way of being and empathy as a communication process or skill. Note that the word *basic* is used in this chapter to modify empathy. The reason is that another form of empathy, advanced empathy, will be discussed in Chapter 9.

EMPATHY AS A WAY OF BEING

A helper cannot communicate an understanding of a client's world without getting in contact with that world. Therefore, a great deal of the discussion on empathy centers on the kind of attending, observing, and listening—the kind of being with—needed to develop an understanding of clients and their worlds. Empathy in this sense is primarily a mode of human contact. Even though it might be metaphysically impossible to actually get inside the world of another person and experience the world as he or she does, it is possible to approximate this. And even an approximation is very useful in helping. The way in which I experience the world differs from the way in which you do, yet there are enough similarities to constitute the basis for mutual empathy. Indeed, some form of empathy for others is essential to caring for them. Caring for clients and their concerns is part of respect. Of course, the kind of caring needed in helping relationships is often tough and action-oriented rather than sentimental.

Rogers (1980) talked about basic empathic listening—being with and understanding—as "an unappreciated way of being" (p. 137) because, despite its usefulness in counseling and therapy, even so-called expert helpers either ignore it or are not skilled in its use. Rogers gave us his description of basic empathic listening, or being with:

> It means entering the private perceptual world of the other and becoming thoroughly at home in it. It involves being sensitive, moment by moment, to the changing felt meanings which flow in this other person, to the fear or rage or tenderness or confusion or whatever that he or she is experiencing. It means temporarily living in the other's life, moving about in it delicately without making judgments. (p. 142)

Indeed, Rogers never ceased to maintain that empathy, respect, and genuineness constituted the foundation of helping.

This ability to enter another's world exacts a price from helpers. They must put themselves and their concerns aside as they listen to and are with their clients. Some wax lyrical about empathy and its place in human commerce: "Empathy, the accepting, confirming, and understanding human echo evoked by the self, is a psychological nutrient without which human life, as we know and cherish it, could not be sustained" (Kohut, 1978, p. 705). Empathy, then, becomes a value, a philosophy, a cause with almost religious overtones. Covey (1989), naming empathic communication one of the seven habits of highly effective people, said that empathy provides psychological air; that is, it helps people breathe more freely in their relationships. Care must be taken, however, not to make a cult out of empathy. However important empathy is, it is not the only mode of being in the helping relationship.

Empathic relationships.   However deep one person's empathic understanding of another, it needs to be communicated to the other. This does not necessarily mean that understanding must always be put into words. Given enough time, people can establish empathic relationships with one another in which understanding is communicated in a variety of rich and subtle ways without necessarily being put into words. A simple glance across a room as one's spouse sees the other trapped in a conversation with a person he or she does not want to be with can communicate worlds of understanding. The glance says: "I know you feel caught. I know you don't want to hurt the other person's feelings. I can feel the struggles going on inside you. I know that you'd like me to rescue you if I can do so tactfully."

People with empathic relationships often express empathy in actions. An arm around the shoulders of someone who has just suffered a defeat can be filled with both support and empathy. I was in the home of a poor family when the father came bursting through the front door shouting, "I got the job!" His wife, without saying a word, went to the refrigerator, got a bottle of beer with a makeshift label on which *CHAMPAGNE* had been written, and offered the bottle to her husband. Beer never tasted so good. Some people do enter caringly into the world of another: this woman communicated her understanding through action, another person might use words, still another person might use both.

Empathic participation in the world of another person obviously admits of degrees. As a helper, you must be able to enter clients' worlds deeply enough to understand their struggles with problem situations or their search for opportunities with enough depth to make your participation in problem management and opportunity development valid and substantial. If your help is based on an incorrect or invalid understanding of the client, then your helping may lead him or her astray. If your understanding is valid but superficial, then you might miss the central issues of the client's life.

Labels — A distortion of understanding.   I remember my reaction to hearing a doctor refer to me as the "hernia in 304." The very labels we learn in our training — paranoid, neurotic, sexual disorder, borderline — can militate against empathic understanding. Helpers forget at times that their labels are interpretations rather than understandings of the client's experience. It is not that certain interpretations of the client's experiences are not sometimes helpful. Rather, problems arise when precast interpretations drawn from our theories about people and their problems preempt our understanding of clients from their points of view. In the process of assigning the correct labels, we may lose the personhood of the client.

EMPATHY AS A COMMUNICATION SKILL

If attending and listening are the skills that enable helpers to get in touch with the world of the client, then empathy is the skill that enables them to communicate their understanding of this world. A secure starting point in helping others is listening to them, struggling to understand their concerns, and sharing this understanding with them. When clients are asked what they find helpful in counseling interviews, understanding gets top ratings. You cannot genuinely respond with understanding to clients unless you are empathic.

Three Dimensions of Communication Skills

The communication skills involved in responding to and engaging in dialogue with clients have three components or dimensions: perceptiveness, know-how, and assertiveness.

Perceptiveness.   Your communication skills are only as good as the accuracy of the perceptions on which they are based.

> Jenny is counseling Frank in a community mental health center. Frank is scared about what is going to happen to him in the counseling process, but he does not talk about it. Jenny senses his discomfort but thinks that he is angry rather than scared. She says: "Frank, I'm wondering what's making you so angry right now." Since Frank does not feel angry, he says nothing. He's startled by what she says and feels even more insecure. Jenny takes Frank's silence as a confirmation of his anger. She tries to get him to talk about it.

Jenny's perception is wrong and disrupts the helping process. The kind of perceptiveness needed to be a good helper is based on the quality of one's being with clients and on practical intelligence. It is developed through experience.

Know-how.   Once you are aware of what kind of response is called for, you need to be able to deliver it. For instance, if you are aware that a client is anxious and confused because this is his first visit to a helper, it does little good if your understanding remains locked up inside you.

> Frank and Jenny end up arguing about what Jenny has called his anger. Frank finally gets up and leaves. Jenny, of course, takes this as a sign that she was right in the first place. Frank goes to see his minister. The minister sees quite clearly that Frank is scared and confused. His perceptions are right. But he lacks the know-how to translate his perceptions into meaningful interactions with him. As Frank talks, the minister nods and says "uh-huh" quite a bit. He is fully present to Frank and listens intently, but he does not know how to respond.

Understandings are lost without the skill of delivering them to the client. Therefore, some delivery technology serves the client's interests.

Assertiveness. Accurate perceptions and excellent know-how are both meaningless unless they are actually used when called for. If you see that a client needs to be challenged and know how to do it but fail to do so, you do not pass the assertiveness test. Your skill remains locked up inside you.

> Edna, a young helper in the Center for Student Development, is in the middle of her second session with Scott, a graduate student. It soon becomes clear to her that he is making sexual overtures. In her training she did quite well in challenging her fellow trainees. The feedback she got from them and the trainer was that she challenged others directly and caringly. But now she feels immobilized. She does not want to hurt Scott or embarrass herself. She tries to ignore his seductive behavior, but Scott takes silence to mean consent.

In this case awareness and know-how are both lost because of a lack of assertiveness. The helper does not act when action is called for. This is not to suggest that assertiveness is an overriding value in and of itself. To be assertive without perceptiveness and know-how is to court disaster.

Examples of Empathy

To get the flavor of things, let's start with a couple of examples. A single, middle-aged man who has been unable to keep a job shares his anger and frustration with a counselor.

CLIENT: I've been to other counselors, and nothing has ever really worked. I don't even know why I'm trying again. But things are so bad. . . . I just have to get a job. I guess something has to be done, so I'm trying it all over again.
HELPER: You're here with mixed feelings. You're not sure that our sessions will help you get a job and keep it, but you feel you have to try something.
CLIENT: Yes, something, but I don't know what the something is. What can I get here that will help me get a job? Or keep a job?

The helper's nondefensive empathic response helps diffuse the client's anger and refocus his energies.

In the next example, a woman who is going to have a hysterectomy is talking with a hospital counselor the night before surgery.

PATIENT: I'm just so afraid. God knows what they'll find when they go in tomorrow. You know, they don't tell you much. I mean they tell you about the procedure, but they don't really tell you about the dangers and the pain.
COUNSELOR: What they don't talk about is the most troubling part of the procedure.
PATIENT: Yes, that's it—not knowing what's going to happen. I think I can take the pain. But the uncertainty drives me up the wall.

The accuracy of the helper's response does not solve the woman's problems, but the patient does move a bit, reducing her anxiety, perhaps, by sharing it.

In the following example a somewhat stressed trainee in a counseling program is talking to her supervisor.

TRAINEE: I don't think I'm going to make a good counselor. The other people in the program seem brighter than I am. Others seem to be picking up the knack of empathy faster than I am. I'm still afraid of responding directly to others, even with empathy. I think I should reevaluate my participation in the program.
SUPERVISOR: You're feeling pretty inadequate and it's getting you down, perhaps even enough to make you begin wondering whether you should be here at all.
TRAINEE: And yet I know that giving up is part of the problem, part of my style. I'm not the brightest, but I'm certainly not dumb either. The way I compare myself to others is not very useful. I know that I've been picking up some of these skills. I do attend and listen well. I'm perceptive even though at times I have a hard time sharing these perceptions with others.

When the trainer hits the mark, the trainee moves forward and explores her problem more fully. Indeed, the function of empathy is to help clients explore their concerns more fully.

Elements of Empathic Understanding and Responding

To begin with, empathy involves translating your understanding of the client's experiences, behaviors, and feelings, as discussed in Chapter 4, into a response through which you share that understanding with the client. If a student comes to you, sits down, looks at the floor, hunches over, and haltingly tells you that he has just failed a test, that his girlfriend has told him she doesn't want to see him anymore, and that he might lose his part-time job because he has been "goofing off," you might respond to him by saying something like this:

HELPER: That's a lot of misery all at once.
CLIENT: I'm not even sure who I'm mad at the most. I think it's myself.

You see that he is both agitated and depressed (affect), and you understand in some initial way what has happened to him (experiences) and what he has done (behaviors) to contribute to the problem situation, and then you communicate to him your understanding of his world. This is basic empathy. If your perceptions are correct — that is, if they reflect the world as the client currently sees it — then it is accurate empathy. In this example, the helper might notice that the client talks about getting angry with the people involved, but does not take much responsibility himself. But at this point the helper "files" this observation because this is not the time to offer it. And, as we see in the example, filing it made sense, because the client himself brings it up.

Another client, after a few sessions spread out over six months, says something like this:

CLIENT (talking in an animated way):   I really think that things couldn't be going better. I'm doing very well at my new job, and my husband isn't just putting up with it. He thinks it's great. He and I are getting along better than ever, even sexually, and I never expected that. We're both working at our marriage. I guess I'm just waiting for the bubble to burst.

HELPER: You feel great because things have been going better than you ever expected — it's almost too good to be true.

CLIENT: What I'm finding out is that I can make things come true instead of sitting around waiting for them to happen as I usually do. I've got to keep making my own luck.

HELPER: You've discovered the secret!

This client, too, talks about her experiences and behaviors and expresses feelings, the flavor of which is captured in the empathic response. The response, capturing as it does both the client's enthusiasm and her lingering fears, is quite useful because the client moves on to her need to make things happen.

The Search for Core Messages

In order to respond with empathy, you can ask yourself, as you listen to the client: "What core message or messages are being expressed in terms of feelings and the experiences and behaviors that underlie these feelings? What is most important in what the client is saying to me?" Once you feel you have identified the core messages, then check out your accuracy with the client. The formula "You feel — — — because — — —" provides the essentials for an empathic response.

Feelings.   "You feel — — —" is to be followed by the correct family of emotions and the correct intensity. For instance, the statements "You feel hurt," "You feel relieved," and "You feel great" specify different families of emotion. The statements "You feel annoyed," "You feel angry," and "You're furious" specify different degrees of intensity in the same family.

Note that the client may actually be talking about feelings experienced in the world, or expressing feelings that arose during the interview, or both. For instance, consider this interchange between a client involved in a child custody proceeding and a counselor.

CLIENT (in a totally relaxed manner):   I get furious with him when he intimates that I'm not a good mother.
HELPER: When you're with him you hate his insinuations. But you seem relaxed about it now.
CLIENT: I'm relaxed because I know I'm a good mother. I'm even more relaxed because my daughter knows I'm a good mother. That's enough for me. . . . Well, the courts have to realize this too.

Helpers need to pick up the difference between the clients' talking *about* their emotions and the emotions they are *expressing* as they talk. Consider another example.

CLIENT (heatedly):   I hate it when I let my fear get the best of me when I'm talking with my father, but it happens all the time! He knows this and he plays me like a harp.
HELPER: You're angry with yourself for letting yourself be had.
CLIENT: What is it about me that I let myself be had? It's not just with my father.

The client felt fear with his father, but now he is angry with himself for letting his fear get the best of him.

Experiences and behaviors.   The "because ———" in the formula is to be followed by an indication of the experiences and behaviors that underlie the client's feelings. "You feel angry *because* he stole your course notes and you let him get away with it" specifies both the client's experience (the theft) and his behavior (in this case, a failure to act) that give rise to the feeling. Some additional examples:

- "You feel sad because moving means leaving all your friends."
- "You feel anxious because the results of the biopsy aren't in yet."
- "You feel frustrated because the social-service bureau keeps making you do things you don't want to do."
- "You feel annoyed with yourself because you didn't reach even the simple goals you set for yourself."
- "You feel hopeless because even your best efforts don't seem to help you lose weight."
- "You feel relieved because sticking to the regime of diet and exercise means that you probably won't need to go on the medicine."
- "You feel angry with me because you see me pushing all the responsibility on you."

Of course, the formula itself is not important; it merely provides a framework for communicating understanding to the client. Though many people in training use the formula to help themselves communicate understanding of core messages, experienced helpers tend to avoid formulas and use whatever wording best communicates their understanding.

Responding to the Context, Not Just the Words

A good empathic response is not based just on the client's immediate words and nonverbal behavior. It takes into account the context of what is said, everything that surrounds and permeates a client's statement. You are listening to clients in the context of their lives. For example, Jeff, a white teenager, is accused of beating a black youth whose car stalled in a white neighborhood. The beaten youth is still in a coma. When Jeff talks to a court-appointed counselor, the counselor listens to what Jeff says in terms of Jeff's upbringing and environment. The context includes the racist attitudes of many people in Jeff's blue-collar neighborhood, the sporadic violence there, the fact that his father died when Jeff was in primary school, a somewhat indulgent mother with a history of alcoholism, and easy access to soft drugs. The following interchange takes place.

JEFF: I don't know why I did it. I just did it, me and these other guys. We'd been drinking a bit and smoking up a bit—but not too much. It was just the whole thing.
COUNSELOR: Looking back, it's almost like it's something that happened rather than something you did, and yet you know, somewhat bitterly, that you actually did it.
JEFF: Bitter's the word! I can be screwed for the rest of my life. It's not like I got up that morning saying that I was going to bash me a black that day.

The counselor's response is in no way an attempt to excuse Jeff's behavior, but it does factor in some of the environmental realities. Later on he will challenge Jeff to decide whether the environment is to own him or whether, to the degree that this is possible, he is to own the environment.

Selective Responding

In most of the examples given so far, the helper has responded to both feelings and the experiences and behaviors underlying the feelings. While this might ordinarily be the best kind of response, helpers can respond selectively so as to emphasize feelings, experiences, or behaviors. Consider the following example of a client who is experiencing stress because of his wife's health and because of concerns at work.

CLIENT: This week I tried to get my wife to see the doctor, but she refused, even though she fainted a couple of times. The kids had no school, so they were underfoot almost constantly. I haven't been able to finish a report my boss expects from me next Monday.

COUNSELOR: It's been an lousy, overwhelming week all the way around.

CLIENT: As bad as they come. When things are lousy both at home and at work, there's no place for me to relax. I just want to get the hell out of the house and find some place to forget it all.

Here the counselor chooses to emphasize the feelings of the client, because she believes that his feelings of frustration and irritation are what is uppermost in his consciousness right now. At another time or with another client, the emphasis might be quite different. In the next example, a young woman is talking about her problems with her father.

CLIENT: My dad yelled at me all the time last year about how I dress. But just last week I heard him telling someone how nice I looked. He yells at my sister about the same things he ignores when my younger brother does them. Sometimes he's really nice with my mother and other times — too much of the time — he's just awful: demanding, grouchy, sarcastic.

COUNSELOR: The inconsistency is killing you.

CLIENT: Right; it's hard for all of us to know where we stand. I hate coming home when I'm not sure which "dad" will be there.

In this response the counselor emphasizes the client's experience of her father, for he feels that this is the core of the client's message. The point is that effective helpers do not stick to rigid formulas in using empathy.

As suggested earlier, fear of intimacy on the part of some clients makes the helping sessions difficult. Since empathy is a kind of intimacy, too much empathy too soon can inhibit rather than facilitate helping. Warmth, closeness, and intimacy are not goals in themselves. The goal of Stage I is to help clients clarify their problems. If empathy, or too much empathy too soon, stands in the way of this goal, then it should be avoided.

Naming and discussing feelings threatens some clients. In this case it might be better to focus on experiences and behaviors and proceed only gradually to a discussion of feelings. The following client, an unmarried man in his mid-thirties who has come to talk about "certain dissatisfactions" in his life, has shown some reluctance to express or even to talk about feelings.

CLIENT (in a pleasant, relaxed voice):   My mother is always trying to make a little kid out of me. And I'm in my mid-thirties! Last week, in front of a group of my friends, she brought out my rubber boots and an umbrella and gave me a little talk on how to dress for bad weather (laughs).

COUNSELOR A: It might be hard to admit it, but I get that down deep you were furious.

CLIENT: Well, I don't know about that. Anyway, at work . . .

COUNSELOR B: So she keeps playing the mother role — to the hilt, it would seem.

CLIENT: And the hilt includes not wanting me to grow up. But I am grown-up . . . well, pretty grown-up.

Counselor A pushes the emotion issue and is met with mild resistance. The client changes the topic. Counselor B, choosing to respond to the "strong mother" issue rather than the more sensitive "being kept a kid and feeling really lousy about it" issue, gives the client more room to move. This works, for the client himself moves toward the more sensitive issue. Some clients are hesitant to talk about certain emotions. One client might find it relatively easy to talk about his anger but not his hurt. For another client it might be just the opposite. Empathy includes the ability to pick up and deal with these differences.

Accurate Empathy: Staying on Track

Although helpers should strive to be accurate in the understanding they communicate, all helpers can be somewhat inaccurate at times. But they can learn from their mistakes. If the helper's response is accurate, the client often tends to confirm its accuracy in two ways. The first is some kind of verbal or nonverbal indication that the helper is right. That is, the client nods, gives some other nonverbal cue, or uses some assenting word or phrase (such as "that's right" or "exactly").

HELPER: So the set up in your neighborhood makes it easy to do things that can get you into trouble.
CLIENT: You bet it does! For instance, everyone's selling drugs. You not only end up using them, but you begin to think about pushing them. It's just too easy.

The second and more substantive way in which clients acknowledge the accuracy of the helper's response is by moving forward in the helping process; for instance, by clarifying the problem situation more fully. In the previous example, the client not only acknowledges the accuracy of the helper's empathy verbally — "you bet it does!" — but, more importantly, outlines the problem situation in greater detail. If the helper again responds with empathy, this leads to the next cycle. The problem situation becomes increasingly clear in terms of specific experiences, behaviors, and feelings.

When a response is inaccurate, the client often lets the counselor know in different ways: he or she may stop dead, fumble around, go off on a new tangent, tell the counselor "That's not exactly what I meant," or even try to provide empathy for the counselor in order to get him or her back on track. A helper who is alert to these cues can get back on track. For example, consider Ben, who lost his wife and daughter in a train crash, as he talks about the changes that have taken place since the accident.

HELPER: So you don't want to do a lot of the things you used to do before the accident. For instance, you don't want to socialize much any more.
BEN (pausing a long time):   Well, I'm not sure that it's a question of wanting to or not. I mean that it takes much more energy to do a lot of things. It takes so much energy for me just to phone others to get together. It takes so much

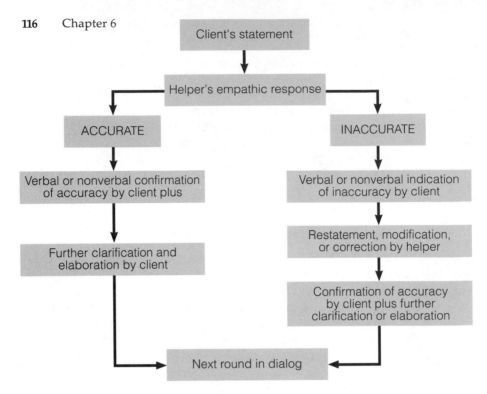

FIGURE 6-1
The Movement Caused by Accurate and Inaccurate Empathy

energy sometimes being with others that I just don't try. It's as if there's a
weight on my soul a lot of the time.

HELPER: It's like a movie of a man in slow motion — it's so hard to do almost
anything.

BEN: Right. I've been in slow motion, low gear in lots of things.

Ben says that it is not a question of motivation but of energy. The difference
is important to him. By picking up on it, the helper gets the interview back
on track. If you are empathic in your relationship with your clients, they will
not be put off by occasional inaccuracies on your part.

Clients are sometimes confused, distracted, and in a highly emotional
state. All these conditions affect the clarity of what they are saying about
themselves. Helpers may fail to pick up what the client is saying because of
the client's confusion or because they themselves have become distracted in
one way or another. In any case, it's a mistake to feign understanding. Genuine helpers admit that they are lost and then work to get back on track
again. A statement like "I think I've lost you. Could we go over that once
more?" indicates that you think it important to stay with the client. It is a
sign of respect. Admitting that you're lost is infinitely preferable to such
cliches as "uh-huh," "ummmm," and "I understand."

Figure 6-1 indicates the movement when helpers are on the mark and
when they fail to grasp what the client has expressed.

THE USES OF EMPATHY

Although some studies suggest that empathy is directly therapeutic in itself (see Burns & Nolen-Hoeksema, 1992) it is certainly not a miracle pill. Empathy can, however, contribute to the success of the overall helping process in a variety of ways. Other things being equal (such as the skill of the helper and the state of the client), empathy can help you accomplish the following goals:

- *Build the relationship.* In interpersonal communication, empathy is a tool of civility. Making an effort to get in touch with another's frame of reference sends a message of respect. Therefore, empathy plays an important part in building a working alliance with clients.
- *Stimulate self-exploration.* Empathy is an unobtrusive tool for helping clients explore themselves and their concerns. When clients are understood, they tend to move on—to explore substantive issues more widely and deeply. Driscoll (1984), in his commonsense way, referred to empathic statements as "nickel-and-dime interventions which each contribute only a smidgen of therapeutic movement, but without which the course of therapeutic progress would be markedly slower" (p. 90).
- *Check understandings.* You may think you understand the client and what he or she has said only to find out, when you share your understanding, that you were off the mark. Therefore, empathy is a perception-checking tool.
- *Provide support.* Since empathy provides a continual trickle of understanding, it is a way of providing support throughout the helping process. It is never wrong to let clients know that you are trying to understand them from their frame of reference.
- *Lubricate communication.* Empathy acts as a kind of communication lubricant; it encourages and facilitates dialogue. It thus encourages collaboration in the helping process.
- *Focus attention.* Empathy helps client and helper alike hold core issues up to the light. This includes core experiences, core behaviors, and core feelings.
- *Restrain the helper.* Empathy keeps helpers from doing useless things such as asking too many questions and giving premature and inept advice. Empathy puts the ball back into the client's court; thus, in its own way, empathy encourages the client to act.
- *Pave the way.* Empathy paves the way for the stronger interventions suggested by the helping model, including challenging a client's assumptions and perceptions, setting goals, formulating strategies, and moving to action.

Empathy as a way of being with others is a human value and needs no justification. Empathy as a communication skill, however, is instrumental; that is, it is good to the degree that achieves the goals outlined here.

What to Do in Communicating Empathy

Here are a few hints to help you improve the quality of your empathic responses.

- *Give yourself time to think.* Beginners sometimes jump in too quickly with an empathic response when the client pauses. "Too quickly" means that they do not give themselves enough time to reflect on what the client has just said in order to identify the core message being communicated. Even the experts pause and allow themselves to assimilate what the client is saying.
- *Use short responses.* I find that the helping process goes best when I engage the client in a dialogue rather than give speeches or allow the client to ramble. In a dialogue the helper's responses can be relatively frequent, but lean and trim. In trying to be accurate, the beginner is often long-winded, especially if he or she waits too long to respond. Again, the question "What is the core of what this person is saying to me?" can help you make your responses short, concrete, and accurate.
- *Gear your response to the client, but remain yourself.* If a client speaks animatedly, telling the helper of his elation over various successes in his life, and she replies accurately, but in a flat, dull voice, her response is not fully empathic. This does not mean that helpers should mimic their clients. It means that part of being with the client is sharing in a reasonable way in his or her emotional tone. Consider this example:

12-YEAR-OLD CLIENT: My teacher started picking on me from the first day of class. I don't fool around more than anyone else in class, but she gets me anytime I do. I think she's picking on me because she doesn't like me. She doesn't yell at Bill Smith, and he acts funnier than I do.

COUNSELOR A: This is a bit perplexing. You wonder why she singles you out for so much discipline.

COUNSELOR B: You're mad because the way she picks on you seems unfair.

It is evident which of these two helpers will probably get through. Counselor A's language is stilted, not in tune with the way a 12-year-old speaks. On the other hand, helpers should not adopt a language that is not their own just to be on the client's wavelength. A white counselor speaking African-American slang sounds ludicrous.

What Not to Do in Attempting to Communicate Empathy

There are a number of things you should *not* do if you are to communicate empathy. Many responses that novice or inept helpers make are poor substitutes for empathy. An example will be used to illustrate a range of poor responses. Peter is a college freshman. This is his second visit to a counselor in the student-services center. After talking about a number of academic concerns, Peter speaks in a halting voice, looking at the floor and clasping his hands tightly between his knees.

PETER: What seems to be really bothering me is a problem with sex. I don't even know whether I'm a man or not, and I'm in college. I don't go out with women. I don't even think I want to. I may . . . well, I may even be gay. . . . Uh, I don't know." (Peter then falls silent and keeps looking at the floor.)

The following are some examples of poor responses to Peter's statements.

- *No response.* It is a mistake to say nothing. Generally, if the client says something significant, respond to it, however briefly. Otherwise the client might think that what he or she has just said doesn't merit a response. Don't leave Peter sitting there in silence.
- *A question.* A counselor might ask something like: "How long has this been bothering you, Peter?" This response ignores what Peter has said and his feelings and focuses rather on the helper's agenda to get more information.
- *A cliché.* A counselor might say: "Many people struggle with sexual identity, Peter, especially at your time of life." This is a cliché. It turns the helper into an instructor and may sound dismissive to the client. The helper is saying, in effect, "You don't really have a problem at all, at least not a serious one." The helper is, once more, involved in his own agenda.
- *An interpretation.* A counselor might say something like this: "My bet is that this sexual issue is probably really just a symptom, Peter. I've got a hunch that you're not really accepting yourself. That's the real problem." The counselor fails to respond to the client's feelings and also distorts the content of the client's communication. The response implies that what is really important is hidden from the client. Once more the helper's agenda takes center stage.
- *Advice.* Another counselor might say: "Stay calm, Peter. Remember that you're struggling with issues that everyone struggles with. I've got a booklet here that I'd like you to read." Advice giving at this stage is out of order; to make things worse, the advice given has a cliché flavor to it. It also distances the helper from Peter's concerns.
- *Parroting.* Empathy is not mere parroting. The mechanical helper corrupts basic empathy by simply restating what the client has said.

COUNSELOR: So it's really a problem with sex that's bothering you, Peter. You're not sure of your manhood even though you're in college. You not only don't go out with women, but you're not even sure you want to. All this makes you think that you might be gay, but you're not sure.

It may be accurate, but it sounds awful. Mere repetition carries no sense of real understanding, of being with the client. A tape recorder could do as well.

▪ *Sympathy and agreement.* Being empathic is not the same as being sympathetic. An expression of sympathy has much more in common with pity, compassion, commiseration, and condolence than with empathic understanding. These are fully human traits, but they are not particularly useful in counseling. Sympathy denotes agreement, whereas empathy denotes understanding and acceptance of the person of the client. At its worst, sympathy is a form of collusion with the client. Note the difference between Counselor A's response to Peter and Counselor B's response.

COUNSELOR A: It's terrible, isn't it Peter, having to struggle with these things and even more terrible to be forced to talk about it.
PETER: Why do these things have to happen to me?
COUNSELOR B: You've got some misgivings about just where you stand with yourself sexually, and it's certainly not easy talking about it.
PETER: I'm having a terrible time talking about it. But I've got to. I've been holding this in for years.

Peter responds to Counselor A with victim talk. Counselor B's response gives Peter the opportunity to deal with his immediate anxiety and then to explore his major concern.

Box 6-1 summarizes some of the main points about the use of empathy as a communication skill.

A Question of Balance

Some helpers expect too much of empathy in and of itself; other helpers do not understand and respect its power. In overly traditional approaches to client-centered helping, empathy is expected to do almost everything; in my mind, too much is expected of empathy. On the other hand, the inability of helpers to communicate empathy diminishes their effectiveness. Understanding clients and their concerns is the rock-solid foundation of helping. As a way of being and a communication skill, empathy should be integrated into a helping model that provides the kind of understanding, challenge, and support that clients need to do their part in establishing a working alliance and moving forward in the problem-management process.

Finally, empathy itself is part of the social-influence process discussed in Chapter 3. Clients say many things and it is impossible to respond to all of them. By necessity helpers are selective. The search for core messages is a selection process. The helper believes that these are core messages, not just for himself or herself, but primarily for the client. But the helper also believes, at some level, that certain messages *should* be important for the client. This involves social influence. On the other hand, empathy is also a stop-

---

BOX 6-1

SUGGESTIONS FOR THE USE OF EMPATHY

1. Remember that empathy is, ideally, a way of being and not just a professional role or communication skill.
2. Attend carefully, both physically and psychologically, and listen to the client's point of view.
3. Try to set your judgments and biases aside for the moment and walk in the shoes of the client.
4. As the client speaks, listen especially for core messages.
5. Listen to both verbal and nonverbal messages and their context.
6. Respond fairly frequently, but briefly, to the client's core messages.
7. Be flexible and tentative enough that the client does not feel pinned down.
8. Use empathy to keep the client focused on important issues.
9. Move gradually toward the exploration of sensitive topics and feelings.
10. After responding with empathy, attend carefully to cues that either confirm or deny the accuracy of your response.
11. Determine whether your empathic responses are helping the client remain focused while developing and clarifying important issues.
12. Note signs of client stress or resistance; try to judge whether these arise because you are inaccurate or because you are too accurate.
13. Keep in mind that the communication skill of empathy, however important, is a tool to help clients see themselves and their problem situations more clearly with a view to managing them more effectively.

---

and-check tool. It makes little sense for the helper to build a picture of clients that clients don't share. Checking perceptions, then, keeps the process honest.

THE ART OF PROBING

In most of the examples used in the discussion of empathy, clients have demonstrated a willingness to explore themselves and their behavior relatively freely. Obviously, this is not always the case. Much as it is essential that helpers respond with empathy to their clients when they do reveal themselves, it is also necessary at times to encourage, prompt, and help clients to explore their concerns when they fail to do so spontaneously. Therefore, the ability to use prompts and probes well is another important skill. Prompts and probes are verbal tactics for helping clients talk about themselves and define their concerns more concretely in terms of specific experiences, behaviors, and feelings and the themes that emerge from an exploration of these. Since the purpose of probes is to help clients explore issues more fully, they are useful in every stage and step of the helping process. They can be used to help clients explore different goals just as they can be used to help them explore initial concerns.

Prompts and probes can take the form of statements, interjections, or questions. Let's look at examples of each.

Statements and Requests That Encourage Clients to Talk and Clarify

Probes do not have to be questions. Counselors can use statements and requests to help clients talk and clarify relevant issues. For example, an involuntary client may come in and then just sit there and fume. The helper probes with a statement:

HELPER: I can see that you're angry, and I have some idea of what it's about, but maybe you could tell me.
CLIENT: Yeah, I bet you don't know. . . . It's you guys. You're so damn smug.

Of their very nature, probing statements make some demand on the client either to talk or to become more specific. They are indirect requests of clients to elaborate on their experiences, behaviors, or feelings.

Consider another example.

HELPER: The Sundays your husband exercises his visiting rights with the children end in his taking verbal pot shots at you, and you get these headaches. I've got a fairly clear picture of what *he* does when he comes over, but it might help if you could describe what you do.
CLIENT: Well, I do nothing.
HELPER: So last Sunday he just began letting you have it for no reason at all. Or just to make you feel bad.
CLIENT: Well . . . not exactly. I asked him about increasing the amount of the child-support payments. And I asked him why he's dragging his feet about getting a better job.

Through probes the counselor helps the client fill in a missing part of the picture. She keeps describing herself as total victim and him as total aggressor and that might not be the full story.

A final example:

HELPER: When the diagnosis of cancer came in two weeks ago, you said that you were both relieved because now you knew and depressed because of what you were going to have to face. You've mentioned that your behavior has been a bit chaotic since then. Tell me what you've been doing.

CLIENT: I've been to the lawyer to get a will drawn up. I'm trying to get all my business affairs in order. I've been writing letters to relatives and friends. I've been on the phone a lot. I guess I've been filling my days with lots of activity to forget about it all.

HELPER: I was wondering how all of that was working for you.

CLIENT: Not too well at all.

In the last example it is clear that the use of probes, like the use of empathy, is part of the social influence process. The helper chooses the topic and asks the client to explore it. Helpers have to keep reminding themselves, however, that the clients' agendas are central, not their own.

Interjections That Help Clients to Focus

A prompt or probe need not be a full question or a statement. It can be a word or phrase that helps focus the client's attention on the discussion.

CLIENT: My son and I have a fairly good working relationship now, even though I'm not entirely satisfied.

HELPER: Not *entirely*.

CLIENT: Well, I should probably say "dissatisfied" because . . .

Another example:

CLIENT: At the end of the day, what with the kids and dinner and cleaning up, I'm not at my best.

HELPER: Meaning that . . .

CLIENT: Meaning that I'm not just tired, but also angry, even hurt — my husband and the kids do practically nothing to help me!

In these cases the probe helps clients say more fully what they were only half saying. The clients name things such as negative emotions that they have not named before. Even such responses as "uh-huh," "mmm," "yes," "I see," "ah," and "oh," as well as nods and the like, can serve as prompts, provided they are used intentionally and are not a sign that the helper's attention is flagging or that he or she does not know what else to do. Interjections can also be nonverbal.

CLIENT (hesitatingly):   I don't know if I can tell you this. I haven't told it to anyone.

HELPER: (The helper says nothing but maintains good eye contact and leans slightly forward.)

CLIENT: Well, my brother had sexual relations with me a few times a couple of years ago. I think about it all the time.

Probes are a form of influence, but they are not ways of extorting from clients things they don't want to give. Statements like, "Oh, come on, tell me! It's really not going to hurt" are forms of extortion.

Questions That Help Clients Talk More Freely and Concretely

I have saved questions until last for a reason. Helpers often ask too many questions. When in doubt as to what to say or do, inept helpers tend to ask questions, as if amassing information were the goal of the helping interview. But questions, judiciously used, can be an important part of your interactions with clients. Here are a few guidelines.

1. Do not ask too many questions.   When clients are asked too many questions, they feel grilled, and this does little for the helping relationship. Many clients instinctively know when questions are just filler, used because the helper does not have anything better to say. I have caught myself asking questions whose answers I didn't even want to know! Let's assume that the helper working with Peter, the young man exploring his sexuality, asks him a whole series of questions: "When did you first feel like this? With whom do you discuss sex? What kinds of sexual experience have you had? What makes you think you are gay?" Peter has every right to say, "Goodbye! no thanks," in response to these intrusions. Helping sessions were never meant to be question-and-answer sessions that go nowhere.

2. Ask questions that serve a purpose.   Don't ask random, aimless questions. Ask questions that have substance to them, questions that help clients get somewhere. If the question is information oriented, it should be information that is useful to the client. Ask questions that challenge the client to think. Consider the following exchange between Peter and his counselor.

PETER: I've got moral principles. I just *can't* be gay. . . . I mean actively gay. I might as well shut down my sex life. You know, forget about it.
COUNSELOR: So your moral principles say no sex. . . . Is that the end of the discussion?
PETER (pausing):   Well, I don't know. Well, no, of course not. Nothing should just end the discussion.

Here the counselor uses a question, not to gather information nor to give advice, but to challenge Peter to consider what he's saying. Later, Peter is talking about his relationship with one of his best friends.

PETER: He has no idea that I'm attracted to him. Just no idea.
COUNSELOR: What's the worst thing that could happen if he knew?

As they talk, the counselor realizes that Peter is making all sorts of assumptions about how people would react to him if they knew he was ambivalent about his sexual identity. The helper uses the question, not to intimate that he should come out of the closet, but to open up the assumptions Peter is making for discussion. The question challenges Peter to discuss these assumptions.

3. Ask open-ended questions that help clients talk about specific experiences, behaviors, and feelings.   As a general rule, ask open-ended questions; that is, questions that require more than a simple yes, no, or similar one-word answer. Not "Now that you've decided to take early retirement, do you have any plans?" but "Now that you've decided to take early retirement, what are your plans?" Counselors who ask closed questions find themselves asking more and more questions. One closed question begets another. Of course, if a specific piece of information is needed, then a closed question may be used. A career counselor might ask: "How many jobs have you had in the past two years?" The information is relevant to helping the client draw up a résumé and a job-search strategy. Open-ended questions help clients fill in things that are being left out of their story, whether experiences, behaviors, or feelings.

CLIENT: He treats me badly, and I don't like it!
HELPER: What does he actually do?
CLIENT: He talks about me behind my back—I know he does. Others tell me
    what he says.

In this example, the helper's question leads to a clearer statement of an experience.

CLIENT: I do funny things that make me feel good.
HELPER: What kinds of things?
CLIENT: I daydream about being a hero, a kind of tragic hero. In my daydreams
    I save the lives of people whom I like but who don't seem to know I exist.
    And then they come running to me but I turn my back on them. I choose to
    be alone! I come up with all sorts of variations of this theme.

In this example the helper's question leads to a clearer outline of the client's internal behaviors.

CLIENT: Since my husband had his stroke, coming home at night is a real
    downer for me.
HELPER: What's this downer like for you?
CLIENT: When I see him sitting immobile in the chair, I'm filled with pity for
    him and the next thing I know it's pity for myself and it's mixed with anger
    or even rage, but I don't know what to be angry at. I don't know how to
    focus my anger. Good God, he's only 42 and I'm only 40!

In this case, the helper's question leads to a fuller description of the client's feelings and emotions. In each of these cases, the client's story gets more specific. Of course, the goal is not just to get more and more detail. Rather it is to get the kind of detail that makes the problem or unused opportunity clear enough to move on to see what can be done about it.

4. In asking questions, keep the focus on the client.   Questions should keep the focus on clients and their interests, not on the theories of helpers. One creative way of doing this is to have clients ask relevant questions of themselves. In the following example, Jill, the helper, and Justin, the client, have been discussing how Justin is letting his impairment — he has lost a leg in a car accident — stand in the way of his picking up his life again. The session has bogged down a bit.

JILL: If you had to ask yourself one question right now, what would it be?
JUSTIN (pausing a long time):   Why are you taking the coward's way out? Why are you just giving up? (His eyes tear up.)

Jill's question puts the ball in Justin's court. It's her way of asking Justin to take responsibility for his part of the session. Justin uses a question to challenge himself in a way he would never have done otherwise. There is no magic to this. Justin might well have said, "I don't know what I'd ask myself. That's your job."

5. Ask questions that help clients get into the stages and steps of the helping model.   In the following example, the client has been talking endlessly about his dissatisfaction with his present job.

COUNSELOR: If you had just the kind of job you wanted, what would it look like?
CLIENT: Well, let me think. . . . It certainly would not be just the money. It would be something that interested me. Well, I'd like a job that gave me freedom to be creative. And I'd like one that . . .

The helper's question not only puts the ball in the client's court, but it asks him to move from merely telling the story to creating the preferred scenario.

A JUDICIOUS MIX OF PROBES AND EMPATHY

Skilled helpers use a mix of probes and empathy to help clients clarify and come to grips with their concerns. Prompts and probes, both verbal and nonverbal, can be overused to the detriment of both the helping relationship and the steps of the helping process. Use probes to help increase rather than decrease the client's initiative. After using a probe to which a client responds, use basic empathy rather than another probe or series of probes as a way of encouraging further exploration. First, if a probe is effective, it will yield information that needs to be listened to and understood. Empathy is called for. Second, empathy, if accurate, tends to place a demand on clients to explore further. It puts the ball back in the client's court. In the following example, the client is a young Chinese-American woman whose father died in China and whose mother is now dying in the United States. She has been talking about the traditional obedience of Chinese women and her fears of slipping into a form of passivity in her American life. She talks about her sister, who gives everything to her husband without looking for anything in return.

COUNSELOR: Is this self-effacing role rooted in your genes?
CLIENT: Well, certainly in my cultural genes. And yet I look around and see many of my American counterparts adopt a different style. A style that frankly appeals to me. But last year, when I took a trip back to China with my mother to meet my half-sisters, the moment I landed I wasn't American. I was totally Chinese again.
COUNSELOR A: What did you learn there?
CLIENT: That I was Chinese!
COUNSELOR B: You learned just how deep the roots go.
CLIENT: And if these roots are so deep, what does that mean for me here? I love my Chinese culture. I want to be Chinese and American at the same time. How to do that, well, I haven't figured that out yet. I thought I had, but I haven't.

Counselor A, instead of responding with empathy to what the client has said, asks a second question, and the client ends up repeating herself more emphatically. Counselor B's use of empathy helps the client move forward.

Both empathy and probing are used throughout the helping process. Both help keep the wheels moving, helping clients explore issues more fully and in a more focused way, take responsibility, say what they want, discover how to get what they want, and move to action. Box 6-2 presents a checklist for effective probing.

THE SHADOW SIDE OF COMMUNICATION SKILLS

One of the reasons I have separated a discussion of the basic communication skills from the steps of the helping process is my feeling that some helpers tend to overidentify the helping process with the communication skills — the tools — that serve it. This is true not only of attending, listening, empathy, and probing, but also of the skills of challenging (that will be discussed in Chapters 8 and 9). Being good at communication skills is not the same as being good at helping. Moreover, an overemphasis on communication skills

---

BOX 6-2

SUGGESTIONS FOR THE USE OF PROBES

1. Keep in mind the goals of probing:
   a. To help nonassertive or reluctant clients tell their stories and engage in other behaviors related to the helping process.
   b. To help clients remain focused on relevant and important issues.
   c. To help clients identify experiences, behaviors, and feelings that give a fuller picture of the issue at hand.
   d. To help clients understand themselves and their problem situations more fully.
2. Use a mix of probing statements, open-ended questions, and interjections.
3. Do not engage clients in question-and-answer sessions.
4. If a probe helps a client reveal relevant information, follow it up with basic empathy rather than another probe.
5. Use whatever judicious mixture of empathy and probing is needed to help clients clarify problems, identify blind spots, develop new scenarios, search for action strategies, formulate plans, and review outcomes of action.
6. Remember that probing is a communication tool that is effective to the degree that it serves the stages and steps of the helping process.

---

can make helping a great deal of talk with very little action. In that case, technique replaces substance. Communication skills are essential, of course, but they still must serve the outcomes of the helping process. If they do nothing more than help you establish a good relationship with a client, then you may be shortchanging that client.

BECOMING PROFICIENT AT COMMUNICATION SKILLS

Understanding communication skills and how they fit into the helping process is one thing. Becoming proficient in their use is another. Trainees think that these "soft" skills should be learned easily and fail to put in the kind of hard work and practice to become fluent in them (Binder, 1990; Georges, 1988). Doing the exercises on communication skills in the student manual that supplements this book and practicing these skills in training groups can help, but these activities are not enough to incorporate empathy into your communication style. Empathy that is trotted out, as it were, for helping encounters is likely to have a hollow ring. These skills must become part of your everyday communication style. After providing some initial training in these skills, I tell students, "Now, go out into your real lives and get good at these skills. I can't do that for you. But you cannot be certified in this program unless and until you demonstrate competency in these skills." Microskills training is just the first step in a process. In the beginning it may be difficult to practice empathy in everyday life, not because it is an impossible skill but because empathy is an improbable event in human communication. Listen to the conversations around you. Using an unobtrusive counter, press the plunger every time you hear empathy used. You may go days without pressing the plunger. In the end, however, you can make empathy a reality in *your* everyday life. And those who interact with you will notice the difference. They probably will not call it empathy. Rather they will say such things as, "She really listens to me" or "He takes me seriously." Empathy, provided that it is not a gimmick, goes far in fostering caring human relationships.

PART THREE

STAGE I OF THE HELPING MODEL AND ADVANCED COMMUNICATIONS SKILLS

The communication skills reviewed in Part 2 are critical tools; with them, you can help clients engage in the process of the helping model. But those skills are not the process itself. Part 3 is a more detailed exposition and illustration of Stage I of the helping model, together with its three steps. The first of those steps is discussed in Chapter 7. Advanced communication skills—those related to helping clients challenge themselves—are discussed and illustrated in Chapters 8 and 9. In Chapter 10, the concept of leverage is further explained.

STAGE I

HELPING CLIENTS IDENTIFY AND CLARIFY PROBLEM SITUATIONS

Clients come to helpers because they need help in managing their lives more effectively. Stage I illustrates three ways in which counselors can help clients understand themselves, their problem situations, and their unused opportunities with a view to managing them more effectively.

Step I-A: Helping clients tell their stories.

Step I-B: Challenging—helping clients overcome blind spots and develop new perspectives on their problem situations and opportunities.

Step I-C: Helping clients discover issues that will make a difference and helping them commit themselves to working on those issues.

These three steps are not restricted to Stage I. They are processes to be used at all stages and steps.

STEP I-A: HELPING CLIENTS TELL THEIR STORIES

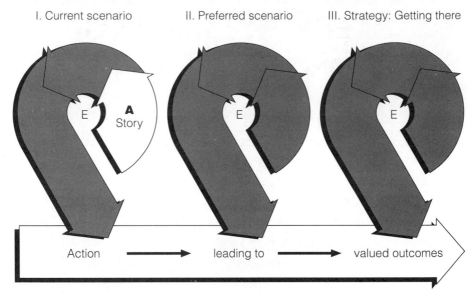

FIGURE 7-1
Step I-A: Helping Clients Tell Their Stories

THE GOALS OF STEP I-A

Figure 7-1 highlights the first step in the helping process. Though "telling the story" may be first in a logical sense, it may not be first in time. As I mentioned in the overview of the helping model, clients do not necessarily start at the beginning. For instance, I once counseled a man who had done all the steps by himself in managing the problems involved with restructuring his life after bypass surgery. He sought my help only because a few things were not going the way he wanted them to. We spent one session fine-tuning his new career goals and action program, and I never saw him again.

Goals are desired outcomes; they are those specific things you are trying to help your clients accomplish. If you are effective, Step I-A will result in clarity, learning, a developing relationship, and an orientation toward action.

• *Clarity.*   Clients will spell out their problem situations and their unused opportunities in terms of specific experiences, behaviors, and feelings. Vague problems lead to vague solutions. The helper's use of the skills of empathy and probing helps clients achieve clarity.
• *Learning.*   Clients will begin to learn things about themselves and their world that will help them manage their lives better. The learnings at this step center around problem situations and opportunities. The client

will be able to answer such questions as: What are my concerns? What needs to be clarified?

■ *Relationship.*   The kind of relationship that serves helping outcomes will develop and strengthen. I don't use the word *deepen* lest client-helper intimacy be seen as an end in itself. Helping is a kind of intimacy, but often it is more like the intimacy experienced by two people working together on a challenging project than the intimacy we associate with friendship or love.

■ *Action.*   Clients will begin to see that some kind of action on their part will be needed if problems are to be managed and opportunities developed. And they will begin to engage in the "little actions" that lead toward problem management and opportunity development.

Since telling the story is not limited to Step I-A, the goals listed here will permeate the entire helping process.

HELPING CLIENTS TELL THEIR STORIES

Self-disclosure — its relationship to personality, its place in relationships, and its role in therapy — has attracted the attention of theoreticians, researchers, and practitioners alike. The entire helping process can be expressed in terms of clients' stories.

> In the course of telling the story of his or her problem, the client provides the therapist with a rough idea of his or her orientation toward life, his or her plans, goals, ambitions, and some idea of the events and pressures surrounding the particular presenting problem. Over time, the therapist must decide whether this problem represents a minor deviation from an otherwise healthy life story. Is this a normal, developmentally appropriate adjustments issue? Or does the therapist detect signs of more thoroughgoing problems in the client's life story? Will therapy play a minor, supportive role to an individual experiencing a low point in his or her life course? If so, the orientation and major themes of the life will be largely unchanged in the therapy experience. But if the trajectory of the life story is problematic in some fundamental way, then more serious, long-term story repair might be indicated. So, from this perspective, part of the work between client and therapist can be seen as life-story elaboration, adjustment, or repair. (Howard, 1991, p. 194)

Using the communication skills outlined in Part Two, counselors help clients engage in self-disclosure or, in the words of the helping model, tell their stories — that is, their concerns about problem situations that are not being managed well and/or opportunities that are not being developed.

In the following example, the client, Louella, is a 34-year-old woman who is separated from her husband and living in public housing in a large city in the Midwest. This is her first session with a social worker.

SOCIAL WORKER: You say that a lot of things are bothering you. What kinds of things?

LOUELLA: Well, they don't take care of this building very well. It's actually dangerous to live here. I've got two children, and I'm always worried about them. I worry about myself, too, but mostly about them.

SOCIAL WORKER: If anything would happen to the kids . . .

LOUELLA: It'd kill me. I think that the elevators scare me the most. . . .

SOCIAL WORKER: They're a real danger spot.

LOUELLA: Often they're broken down. But that's not the real scary part. . . . People that don't live here can get in, and women have been raped. . . . I'm afraid the kids could get hurt. That's the real scary part. . . . And no one seems to do anything about it. This building's not at the top of anyone's list, I can tell you that. I don't know whether to stay here or go, but I don't know where I'd go.

SOCIAL WORKER: What makes things worse is that no one seems to care and you feel you might be stuck here — though you're not sure.

Here attending, listening, empathy, and probing are all used to build a relationship with the client and help her tell her story. The story deals mostly with conditions that are beyond the client's personal control. Could the place be safer? Does Louella have other options? These questions remain to be answered.

Some clients are highly verbal and quite willing to tell almost everything at the first sitting. Take the case of Martin.

Martin, 24, asks a counselor in private practice for an appointment in order to discuss "a number of issues." Martin is both verbal and willing to talk, and his story comes tumbling out in rich detail. While the helper uses the skills of attending, listening, empathy, and probing, she does so sparingly. Martin is too eager to tell his story.

Although trained as a counselor, Martin is currently working in his uncle's business because of an offer he "could not turn down." He is doing very well, but he feels guilty because service to others has always been a value for him. And, although he likes his current job, he also feels hemmed in by it. A year of study in Europe has whetted his appetite for "adventure." He feels that he is nowhere near fulfilling the "great expectations" he has for himself.

He also talks about his problems with his family. His father is dead. He has never gotten along well with his mother, and now that he has moved out of the house, he feels that he has abandoned his younger brother, whom he loves very much.

Martin is also concerned about his relationship with a woman who is two years older than he. They have been involved with each other for about three years. She wants to get married, but he feels that he is not ready. He still has "too many things to do" and would like to keep the arrangement they have. He also admits that he is "a bit promiscuous."

This whole complex story comes tumbling out in a torrent of words. Martin feels free to skip from one topic to another. The way Martin tells his story is part of his impulsive style. At one point he stops, smiles, and says, "Boy, I didn't realize that I had so much on my plate!" While all of these issues bother him, he does not feel overwhelmed by them.

Consider a different example. Terry, a young woman in her late teens who has been arrested several times for street prostitution, is an involuntary client. She ran away from home when she was 16 and is now living in a large city on the East Coast. Like many other runaways, she was seduced into prostitution "by a halfway decent pimp." Now she is pretty street smart. She is also cynical about herself, her occupation, and the world. She is forced to see a counselor as part of the probation process. As might be expected, Terry is quite hostile during the interview. She has seen other counselors and sees the interview as a game. The counselor already knows a great deal of the story because of the court record.

TERRY: If you think I'm going to talk with you, you've got another think coming. What I do outside is my business.
COUNSELOR: You don't have to talk about what you do outside. We could talk about what we're doing here in this meeting.
TERRY: I'm here because I have to be. You're here either because you're dumb and like to do stupid things or because you couldn't get a better job. You people are no better than the people on the street. You're just more "respectable." That's a laugh.
COUNSELOR: So nobody could really be interested in you.
TERRY: I'm not even interested in me!
COUNSELOR: And if you're not interested in yourself, then no one else could be, including me.

The counselor tries to help Terry tell, not the story "out there" — the story of the lifestyle that gets her in trouble — but the story that both of them are living right now "in here." Even though there are some explosive moments, Terry goes on to tell a story of her distrust of just about everyone, especially anyone connected with "the system." She emphasizes her contempt for the kind of people who try to "help" her.

Each client is different and will approach the telling of the story in different ways. Some of the stories will be long and detailed; others, short and stark. Some will be filled with emotion; others will be told coldly, even though they are stories of terror. Some stories will be, at least at first blush, single-issue stories — "I want to get rid of these headaches" — others, like Martin's, will be multiple-issue stories. Some will be told readily, tumbling out with practically no help; others will be told only haltingly or grudgingly. Some clients will tell the core story immediately; others will tell a secondary story first in order to test the helper's reactions. Some stories will deal with the inner world of the client almost exclusively — "I hate myself," "I hear voices" — others will deal with the outer world, problems in the social settings of life. Still others will deal with a combination of inner and outer concerns. In all these cases the job of helpers is to be with their clients in such fully human and skillful ways that the clients are helped to tell their stories. A story that is brought out in the open is the starting point for possible constructive change. Often the very airing of the story helps the client see it more objectively.

When clients like Martin pour out their stories, you may let them go on or you may insist on some kind of dialogue. If the client tells the "whole" story in a more or less nonstop fashion it will be impossible for you to respond empathetically all at once to everything the client has said. But you can then help the client review the most salient issues in some orderly way. Expressions like the following can be used to help the client review the core parts of the story:

- "There's a lot of pain and confusion in what you've just told me. Let's see if I've understood some of the main points you've made. First of all . . ."
- "You've said quite a bit. Let's see if we can see how all the pieces fit together. The main issue seems to be . . ."

At this point your empathy will let the client know that you have been listening intently and that you are concerned about him or her.

Encouraging Clarity

Clarity means the discussion of problem situations and unused opportunities in terms of specific experiences, specific behaviors, and specific feelings. Vagueness and ambiguity lead nowhere. It is up to the counselor to help the client move to clarity. Attending, listening, empathy, and probing are the tools counselors use to help clients achieve action-oriented clarity.

Janice's husband has been suffering from severe depression for over a year. All the attention is on him. One day, after a fainting spell, she too talks with a counselor. At first, feeling guilty about her husband, she is hesitant to discuss her own problems. In the beginning she says only that her social life is "a bit restricted." With the help of empathy and probing on the part of the helper, her story emerges. "A bit restricted" turns, bit by bit, into the following much fuller story. This is a summary. Janice did not say this all at once.

It's like nothing I've ever experienced. I move from guilt to anger to indifference to hope to despair. I have no social life. Friends avoid us because it is so difficult being with John. I feel I shouldn't leave him, so I stay at home. He's always tired, so we have little interaction. I feel like a prisoner in my own home. Then I think of the burden he's carrying and the roller-coaster emotions start all over again. Sometimes I can't sleep, then I'm as tired as he. He is always saying how hopeless things are and even though I don't feel like he does, some kind of hopelessness creeps into my bones. I feel that a stronger woman, a more selfless woman, a smarter woman would find ways to deal with all of this. But I end up feeling that I'm not dealing with it at all. From day

> to day I think I cover most of this up, so he doesn't see it nor do the few people who come around see it. I'm as alone as he is.

Here, then, is a whole series of negative experiences and emotions. The actions she takes—staying at home, covering her feelings up—are part of the problem, not the solution. But now that the story is out in the open, there is some possibility of doing something about it.

In another case, a bulimic woman, now under psychiatric care, says that she acted "a little erratically at times" with some of her classmates in law school. The counselor, using empathy and probes, helps her tell her story in much greater detail. Like Janice, she does not say all of this at once, but this is the fuller picture of "a little erratically."

> I usually think about myself as plain even though when I take care of myself some say that I don't look that bad. Ever since I was a teenager, I've preferred to go it alone, because it was safer. No fuss, no muss, and especially no rejection. In law school, right from the beginning I entertained romantic fantasies about some of my classmates who I didn't think would give me a second look. I pretended to have meals with those who attracted me and then I'd have fantasies of having sex with them. Then I'd purge, getting rid of the fat I got from eating and getting rid of the guilt. In my mind I picked them off one at a time. I would go out of my way to meet my latest imagined sex partner. And then I'd be rude to him to get back at him for what he did to me. That was my way of getting rid of him. He wouldn't know what hit him.

She was not really delusional, but gradually—with a kind of twisted logic— her external behaviors began to reinforce her internal fantasizing behavior. However, once her story became "public"—that is, reviewed by both herself and her helper—she began to take back control of her life.

The Pluses and Minuses of Discussing the Past

Some schools of psychology suggest that problem situations are not clear and fully understood until they are understood in terms of their roots. However, the causes of current behavior, especially the remote causes, are seldom evident. Long ago Deutsch (1954) noted that it is often almost impossible, even in carefully controlled laboratory situations, to determine whether Event B, which follows Event A in time, is actually caused by Event A. Asking clients to come up with such causes—"Why did your father reject you?"—is often inviting them to whistle in the wind. Consider the case of a man who is talking to a counselor because he is being sued by some of his employees for unfair practices. He describes his current style with people and some of the problems it has created both in the workplace and at home.

HELPER: Do you think your father played a part in this?
CLIENT: I think my relationship with my father has a lot to do with the way I
     am now. He was so harsh with me and so now I think I'm going the oppo-

site way and being too soft with the kids. He never really told me that I did anything good. I think that affects the way I feel about myself today. And I could tell you lots of stories about how he treated my mother. I don't want to say that he was mean. At least he was not physically cruel. There are certain things he wouldn't let her do. There is some of that in me, but I think it has a different twist. Actually, I think it's very complex. Why would I be too soft at home and supposedly too hard at work? Well, let me go back to my father . . .

Such talk is a bottomless pit. It deals mostly with what happened to clients (their experiences) rather than what they did about what happened to them (their behaviors). This is not to say that a person's past does not influence current behavior. Nor does it imply that a client's past should not be discussed.

Talking about the past to make sense of the present.   There are ways of talking about the past that help clients make sense of the present (see Driscoll, 1984). But making sense of the present needs to remain center stage. Thus, how the past is discussed is more important than whether it is discussed. The following man has been discussing how his "interpersonal style" gets him into trouble.

CLIENT: Until we began talking I had no idea about how alive and well my father is — in me. For instance, even though I hated his cruelty, it lives on in me in much smaller ways. He beat my brother. I just cut him down to size verbally. He told my mother what she could do and couldn't do. I try to get my mother to adopt my set of living arrangements for her own good. There's a whole pattern that I haven't noticed. I've inherited more than his genes.

In sum, if the past can add clarity to current experiences, behaviors, and emotions, then let it be discussed. If it can provide hints as to how self-defeating thinking and behaving can be changed now, let it be discussed. The past, however, should never become the principal focus of the client's self-exploration. When it does, helping tends to "lose the name of action." Note how different this client's discussion of the past is from the client cited earlier. The previous client, with the complicity of the helper, was wallowing in the past. This client, in reviewing the past, learns from it.

Talking about the past to be liberated from it.   A potentially dangerous logic can underlie discussions of the past. It goes something like this: "I am what I am today because of my past; I cannot change my past. So how can I be expected to change today?"

CLIENT: I was all right until I was about thirteen. I began to dislike myself as a teenager. I hated all the changes — the awkwardness, the different emotions. I was so impressionable. I began to think that life actually must get worse and worse instead of better and better. I just got locked into that way of thinking.

This is not liberation talk. The past is still casting its spell. The following case, of the father of a boy who has been sexually abused by a minister of their church, is quite different. Dealing with the incident helps him review his own abuse by his father. In a tearful session he tells the whole story. In a second interview he has this to say.

CLIENT: Someone said that good things can come from evil things. What happened to my son was evil. But we'll give him all the support he needs. I kept it all in until now. It was all locked up inside. I was so ashamed and my shame became part of me. Last week was like throwing off a dirty cloak that I'd been wearing for years. Getting it out was so painful and now I feel so different, so good. I wonder why I had to hold it in for so long.

This is liberation talk. When counselors help or encourage clients to talk about the past, they should have a clear idea of what their objective is. Is it to learn from it? Is it to be liberated from it? To assume that there is some "silver bullet" in the past that will solve everything now is a search for magic.

ASSESSMENT AND LEARNING

Stage I in the helping process can be seen as the assessment stage — finding out or rather learning what's wrong, what resources are not being used, what opportunities lie fallow. Recall that learning takes place when options are identified, increased, and acted on. Assessment makes sense to the degree that it contributes to learning, to increasing the client's options. Unfortunately, although a great deal of work has been poured into assessment methods such as psychological tests, very little has been devoted to determining what payoff these methods have for clients. Client-centered assessment is the ability to understand clients, to spot "what's going on" with them, to see what they do not see and need to see, to make sense out of their chaotic behavior and help them make sense out of it — all at the service of helping them manage their lives and develop their resources more effectively. Assessment, then, is not something helpers do to clients: "Now that I have my secret information about you, I can fix you." Rather, it is a kind of learning in which both client and helper participate.

Assessment is a way of asking, What's really going on here? It takes place, in part, through the kind of reality-testing listening discussed in Chapter 5. As helpers listen to clients, they hear more than the client's point of view. They listen to the client's deep conviction that he is "not an alcoholic," and at the same time they see the trembling hand, smell the alcohol-laden breath, and hear the desperate tone of voice. The purpose of such listening is not to place the client in the proper diagnostic category. The purpose is to be open to any kind of information or understanding that will enable them to help the client.

In medicine, assessment is usually a separate phase. In helping there is an interplay between assessment and intervention. Helpers who continually listen to clients in context are engaging in a kind of ongoing assessment that is in keeping with the clients' interests. Assessment is thus part and parcel of all stages and steps of the helping model. However, as in medicine, some initial assessment of the seriousness of the clients' concerns is called for.

Initial Assessment

Clients come to helpers with problems of every degree of severity. Problems run, objectively, from the inconsequential to the life-threatening. Subjectively, however, even a relatively inconsequential problem can be experienced as severe by a client. If a client thinks that a problem is critical, even though by objective standards the problem does not seem to be that bad, then for him or her it is critical. In this case, the client's tendency to "catastrophize," to judge a problem situation to be more severe than it actually is, itself becomes an important part of the problem situation. One of the tasks of the counselor in this case will be to help the client put the problem in perspective or to teach the client how to distinguish between degrees of problem severity.

Mehrabian and Reed (1969) suggested the following formula as a way of determining the severity of any given problem situation:

$$\text{Severity} = \text{Distress} \times \text{Uncontrollability} \times \text{Frequency}$$

The multiplication signs in the formula indicate that these factors are not just additive. Even low-level anxiety, if it is uncontrollable or persistent, can constitute a severe problem; that is, it can severely interfere with the quality of a client's life.

One way of viewing helping is to see it as a process in which clients are helped to control the severity of their problems in living. The severity of any given problem situation will be reduced if the stress can be reduced, if the frequency of the problem situation can be lessened, or if the client's control over the problem situation can be increased. Consider the following example.

Indira is greatly distressed because she experiences migraine-like headaches, sometimes two or three times a week, and seems to be able to do little about them. No painkillers seem to work. She has even been tempted to try strong narcotics to control the pain, but she fears she might become an addict. She feels trapped. For her the problem is quite severe because stress, uncontrollability, and frequency are all high.

Eventually Indira is referred to a doctor who specializes in treating headaches. He is an expert in both medicine and human behavior. He first helps her to see that the headaches are getting worse because of her tendency to "catastrophize" whenever she experiences one. That is, the self-talk she engages in ("How intolerable this is!") and the way she fights the headache actually add to its severity. He helps her control her stress-inducing self-talk and teaches her relaxation techniques. Neither of these gets rid of the headaches, but they help reduce the degree of stress she feels.

Second, he helps her identify the situations in which the headaches seem to develop. They tend to come at times when she is not managing other forms of stress well, as when she lets herself become overloaded and gets behind at work. He helps her see that, while her headaches constitute a very real problem, they are part of a larger problem situation. Once she begins to control and reduce other forms of stress in her life, the frequency of the headaches begins to diminish.

Third, the doctor also helps Indira spot cues that a headache is beginning to develop. She finds that, while she can do little to control the headache once it is in full swing, there are things she can do during the beginning stage. For instance, she learns that medicine that has no effect when the headache is in full force does help if it is taken as soon as the symptoms appear. Relaxation techniques are also helpful if they are used soon enough. Indira's headaches do not disappear completely, but they are no longer the central reality of her life.

The doctor helps the client manage a severe problem situation much better than she has been by helping her reduce stress and frequency, while increasing control. In other words, he helps the client learn.

Assessing and Promoting Clients' Abilities to Learn

William Miller (1986) talked about one of the worst days of his life. Everything was going wrong at work: projects were not working out, people were not responding, the work overload was bad and getting worse—nothing but failure all around. Later that day, over a cup of coffee, he took some paper, put the title "Lessons Learned and Relearned" at the top and put down as many entries as he could. Some hours and seven pages later, he had listed 27 lessons. The day turned out to be one of the best of his life. Both his attitude and his ability to come up with solutions to his problems helped so much that he began writing a "Lessons Learned" journal every day. He learned not to get caught up in self-blame and defeatism. Subsequently, on days when things were not working out, he would say to himself, "Ah, this will be a day filled with learnings!" In Stage I, indeed throughout the helping process, assessment can be seen as this search for learnings.

Assessment as the Search for Resources

Unskilled or new helpers concentrate on clients' deficits. Skilled, experienced helpers, as they listen to and observe clients, do not blind themselves to deficits, but they are quick to spot clients' resources, whether used, unused, or even abused. These resources can become the building blocks for the future. For instance, Terry, the street prostitute mentioned earlier, had obvious deficits. She is engaged in a dangerous and self-defeating lifestyle. But as the probation counselor listened to Terry, he spotted many different resources. Terry was a tough, street-smart woman. The very virulence of her cynicism and self-hate, the very strength of her resistance to help, and her almost unchallengeable determination to go it alone were all signs of re- sources. Many of her resources were being used in self-defeating ways, but they were still resources.

Helpers need a resource-oriented mind-set in all their interactions with clients.

CLIENT: I practically never stand up for my rights. If I disagree with what any-
one is saying—especially in a group—I keep my mouth shut.
COUNSELOR: And when you do speak up?
CLIENT (pausing):   I suppose that on the rare occasions when I do speak up,
the world doesn't fall in on me. Sometimes others do actually listen to me.
But I still don't seem to have much impact on anyone.
COUNSELOR A: So speaking up, even when it's safe, doesn't get you much.
COUNSELOR B: So when you do speak up, you don't get blasted, you even get
a hearing. Tell me what makes you think you don't exercise much influence
when you speak.

Counselor A misses the resource mentioned by the client. Although it is true that the client habitually fails to speak up, he does have some impact when he does speak—others do listen, at least sometimes—and this is a resource. Counselor A emphasizes the deficit. Counselor B notes the asset.

The search for resources is especially important when the story being told is bleak. I once listened to a man's story that included a number of bone-jarring life defeats—a bitter divorce, false accusations that led to his dis-missal from a job, months of unemployment, serious health concerns, months of homelessness, and more. The only emotion the man exhibited during his story was depression. Toward the end of the session, we had this interchange.

HELPER: Just one blow after another, grinding you down.
CLIENT: Grinding me down, and almost doing me in.
HELPER: Tell me a little more about the "almost" part of it.
CLIENT: Well, I'm still alive, still sitting here talking to you.
HELPER: Despite all these blows, you haven't fallen apart. That seems to say
something about the fiber in you.

At the word *fiber*, the client looked up, and there seemed to be a glimmer of something else besides depression in his face. I put a line down the center of a newsprint pad. On the left side I listed the life blows the man had experi-enced. On the right I put the word *fiber*. Then I said, "Let's see if we can add to the list on the right side." We came up with a list of the man's resources— his musical talent, his honesty, his concern for and ability to talk to others, and so forth. After about a half-hour, he smiled — weakly, but he did smile. He said that it was the first time he could remember smiling in months.

RELUCTANT AND RESISTANT CLIENTS

It is impossible to be in the business of helping people for long without encountering both reluctance and resistance (Clark, 1991; Ellis, 1985; Fremont & Anderson, 1986, 1988; Friedlander & Schwartz, 1985; Otani, 1989). The differences between reluctance and resistance are spelled out here. In practice a mixture of both is often found in the same client.

Reluctance — Misgivings about Change

Reluctance — some of which is found in everyone — refers to the ambiguity clients feel when they know that managing their lives better is going to exact a price. Clients are not sure whether they want to pay that price. The incentives for not changing drive out the incentives for changing. This accounts for the sad record for discretionary change mentioned earlier. Helping clients manage their reluctance is a case-management issue, but reluctance is so prevalent and so important that it needs to be dealt with up front.

Reluctant clients.   Clients exercise reluctance in many different, often covert, ways. They talk about only safe or low-priority issues, seem unsure of what they want, benignly sabotage the helping process by being overly cooperative, set unrealistic goals and then use them as an excuse for not working, don't work very hard at changing their behavior, and are slow to take responsibility for themselves. They tend to blame others or the social settings and systems of their lives for their troubles and play games with helpers. Reluctance admits of degrees; clients come "armored" against change to a greater or lesser degree.

Some of the roots of reluctance.   The reasons for reluctance are many. They are built into our humanity, plagued as it is by the "psychopathology of the average." Here is a sampling.

Fear of intensity.   If the counselor uses high levels of attending, accurate empathy, respect, concreteness, and genuineness, and if the client cooperates by exploring the feelings, experiences, and behaviors related to the problematic areas of his or her life, the helping process can be an intense one. This intensity can cause both helper and client to back off. Skilled helpers know that counseling is potentially intense. They are prepared for it and know how to support a client who is not used to such intensity.

Lack of trust.   Some clients find it very difficult to trust anyone, even a most trustworthy helper. They have irrational fears of being betrayed. Even when confidentiality is an explicit part of the client-helper contract, some clients are very slow to reveal themselves.

Fear of disorganization.   Some people fear self-disclosure because they feel that they cannot face what they might find out about themselves. The client feels that the facade he or she has constructed, no matter how much energy must be expended to keep it propped up, is still less burdensome than exploring the unknown. Such clients often begin well, but retreat once they begin to be overwhelmed by the data produced in the problem-exploration process. Digging into one's inadequacies always leads to a certain amount of disequilibrium, disorganization, and crisis. But, just as disequilibrium for the child is the price of growth, so for the adult growth often takes place at crisis points. High disorganization immobilizes the client; very low disorganization is often indicative of a failure to get at the central issues of the problem situation.

Shame.   Shame is a much overlooked experiential variable in human living (Egan, 1970; Kaufman, 1989; Lynd, 1958). It is an important part of disorganization and crisis. The root meaning of the verb *to shame* is "to uncover, to expose, to wound," a meaning that suggests the process of painful self-exploration. Shame is not just being painfully exposed to another; it is primarily an exposure of self to oneself. In shame experiences, particularly sensitive and vulnerable aspects of the self are exposed, especially to one's own eyes. Shame is often sudden: in a flash, and without being ready for such a revelation, one sees heretofore unrecognized inadequacies. Shame is sometimes touched off by external incidents, such as a casual remark someone makes. It could not be touched off by such insignificant incidents unless, deep down, one was already ashamed. A shame experience might be defined as an acute emotional awareness of a failure to *be* in some way.

Fear of change.   The title of an article written by Bugental and Bugental in 1986 said it all: "A fate worse than death: The fear of changing." Some people are afraid of taking stock of themselves because they know, however subconsciously, that if they do, they will have to change — that is, surrender comfortable but unproductive patterns of living, work more diligently, suffer the pain of loss, acquire skills needed to live more effectively, and so on. For instance, a husband and wife may realize, at some level of their being, that if they see a counselor they will have to reveal themselves and that, once the cards are on the table, they will have to go through the agony of changing their style of relating to each other.

In a counseling group I once dealt with a man in his sixties who complained about constant anxiety that was making his life quite painful. He told the story of how he had been treated brutally by his father until he finally ran away from home. We have already seen his logic earlier in this chapter: "No one who grows up with scars like mine can be expected to take charge of his life and live responsibly." He had been using his mistreatment as a youth as an excuse to act irresponsibly at work (he had an extremely poor job record), in his relationship with himself (he drank excessively), and in his marriage (he had been uncooperative and unfaithful and yet expected his wife to support him). The idea that he could change, that he could take responsibility for himself even at his age, frightened him, and he wanted to run away from the group. But, since his anxiety was so painful, he stayed. And he changed.

Each and every one of us needs only to look at our own struggles with growth, development, and maturity to add to this list of the roots of reluctance.

Resistance — Fighting Coercion

Reluctance is often passive. Resistance is active. Resistance refers to the reaction of clients who in some way feel coerced. It is their way of fighting back (*see* Dimond, Havens, & Jones, 1978; Driscoll, 1984). For instance, spouses who feel forced to come to marriage counseling sessions are often resistant. Resistance is the client's reaction to a power play. It may well arise because of a misperception: the client may see coercion where it does not exist. But, since people act on their perceptions, the result is still some form of fighting back. Resistant clients are likely to present themselves as not needing help, to feel abused, to show no willingness to establish a relationship with the helper, to con helpers whenever possible. They may be resentful, make active attempts to sabotage the helping process, terminate the process at the earliest possible moment, and be either testy or actually abusive and belligerent. In marriage counseling, one spouse might be there willingly and the other, because he or she feels pressured by the helper, the spouse, or both. Resistance to helping is, of course, a matter of degree, and not all these behaviors in their most virulent forms are seen in all resistant clients.

Involuntary clients — mandated clients — are often resisters. Helpers can expect to find a large proportion of such clients in settings where clients are forced to see a counselor (a high school student in trouble with a teacher must go to a helper as a form of punishment) or in settings where a reward can be achieved only on the condition of being involved in some kind of counseling process (going to a counselor is the price that must be paid to get a certain job). Clients like these are found in schools, especially schools below college level; in correctional settings; in marriage counseling, especially if it is court-mandated; in court-related settings; and in employment, welfare, and other social agencies.

There are all sorts of reasons for resisting; clients can experience coercion in a wide variety of ways. The following kinds of clients are likely to be resistant:

- Clients who see no reason for going to the helper in the first place.
- Clients who resent third-party referrers (parents, teachers, agencies) and whose resentment carries over to the helper.
- Clients in medical settings who are asked to participate in counseling.
- Clients who feel awkward in participating, who do not know how to be "good" clients.
- Clients who have a history of being rebels.

- Clients who see the goals of the helper or the helping system as different from their own. For instance, the goal of counseling in a welfare setting may be to help the clients become financially independent, whereas some clients might be satisfied with the present arrangement. The goal of helping in a mental hospital may be to help the clients get out, whereas some clients might feel quite comfortable in the hospital.
- Clients who have developed negative attitudes about helping and helping agencies and who harbor suspicions about helping and helpers. Helpers are referred to in derogatory and inexact terms (as, for example, "shrinks").
- Clients who believe that going to a helper is the same as admitting weakness, failure, and inadequacy. They feel that they will lose face by going. By resisting the process, they preserve their self-esteem.
- Clients who feel that counseling is something that is being done to them. They feel that their rights are not being respected.
- Clients who feel that they have not been invited by helpers to be participants in the decisions that are to affect their lives. This includes expectations for change and decisions about procedures to be used in the helping process.
- Clients who feel a need for personal power and find it through resisting a powerful figure or agency: "I may be relatively powerless, but I still have the power to resist."
- Clients who distrust helpers and need to test their level of support or their competence.
- Clients who dislike their helpers, but do not discuss their dislike with them.
- Clients who have a conception of the degree of change desired that differs from the helper's conception.

All of these clients have in common a feeling of being coerced to engage in the process, even if they are not sure precisely of the source of the coercion. Resistance, of course, can be a healthy sign. Clients are fighting back.

Many socio-psychological variables — sex, prejudice, race, religion, social class, upbringing, cultural and subcultural blueprints, and the like — can play a part in resistance. For instance, a man might subconsciously resist being helped by a woman, and vice versa. A black person might subconsciously resist being helped by a white, and vice versa. A person with no religious affiliation might think that helping coming from a minister will be "pious" or will automatically include some form of proselytizing.

Helping Clients Manage Reluctance and Resistance

Because reluctance and resistance are such pervasive phenomena, helping clients manage them is part and parcel of all of our interactions with clients (Kottler, 1992).

Unhelpful responses to reluctance and resistance.   Helpers, especially beginning helpers who are unaware of the pervasiveness of reluctance and resistance, are often disconcerted by them. They may feel confused, panicked, irritated, hostile, guilty, hurt, rejected, meek, or depressed. Distracted by these unexpected feelings, they react in any of several unhelpful ways.

- They accept their guilt and try to placate the client.
- They become impatient and hostile and manifest these feelings either verbally or nonverbally.
- They do nothing, in the hope that the reluctance or resistance will disappear.
- They lower their expectations of themselves and proceed with the helping process, but in a half-hearted way.
- They try to become warmer and more accepting, hoping to win the client over by love.
- They blame the client and end up in a power struggle with him or her.
- They allow themselves to be abused by clients, playing the role of a scapegoat.
- They lower their expectations of what can be achieved by counseling.
- They hand the direction of the helping process over to the client.
- They give up and terminate counseling.

In a word, when helpers engage "difficult" clients, they experience stress, and some give in to self-defeating fight-or-flight approaches to handling it.

The source of this stress is not just the behavior of clients; it also comes from the helper's own self-defeating attitudes and assumptions about the helping process. Here are some of these attitudes and assumptions.

- All clients should be self-referred and adequately committed to change before appearing at my door.
- Every client must like me and trust me.
- I am a consultant and not a social influencer; it should not be necessary to place demands on clients or even help them place demands on themselves.
- Every unwilling client can be helped.
- No unwilling client can be helped.
- I alone am responsible for what happens to this client.
- I have to succeed completely with every client.

Effective helpers neither court reluctance and resistance nor are surprised by it.

More productive approaches to dealing with reluctance and resistance.
In a book like this it is impossible to identify every possible form of reluctance and resistance, much less provide a set of strategies for managing

each. Here are some principles and a general approach to managing reluctance and resistance in whatever form they take.

See some reluctance and resistance as normative.   Help clients see that they are not odd because they are reluctant. Beyond that, help them see the positive side of resistance. It may well be a sign of their affirmation of self.

See reluctance as avoidance.   Reluctance is a form of avoidance that is not necessarily tied to client ill will. Therefore, you need to understand the principles and mechanisms underlying avoidance behavior, which is often discussed in texts dealing with the principles of behavior (*see* Watson & Tharp, 1993). Some clients avoid counseling or give themselves only halfheartedly to it because they see it as lacking in suitable rewards or even punishing. If this is the case, then counselors have to help them in their search for suitable rewards. Constructive change is more rewarding than the status quo but this might not be the perception of any given client.

Explore your own reluctance and resistance.   Examine reluctance and resistance in your own life. How do you react when you feel coerced? How do you resist personal growth and development? If you are in touch with the various forms of reluctance and resistance in yourself and are finding ways of overcoming them, you are more likely to help clients deal with theirs.

Examine the quality of your interventions.   Without setting off on a guilt trip, examine your helping behavior. Ask yourself whether you are doing anything to coerce clients, thereby running the risk of provoking resistance. For example, you may have become too directive without realizing it. Furthermore, take stock of the emotions that well up in you when clients lash back or drag their feet and the ways in which these emotions might be leaking out. Do not deny these feelings. Rather, own them and find ways of coming to terms with them. Do not overpersonalize what the client says and does. If you are allowing a hostile client to get under your skin, you are probably reducing your effectiveness.

Accept and work with the client's reluctance and resistance.   This is a central principle. Start with the frame of reference of the client. Accept both the client and his or her reluctance/resistance. Do not ignore it or be intimidated by it. Let clients know how you experience it and then explore it with them. Model openness to challenge. Be willing to explore your own negative feelings. The skill of direct, mutual talk, called immediacy, (and discussed in Chapter 9), is extremely important here. Help clients work through the emotions associated with reluctance/resistance. Avoid moralizing. *Befriend* the reluctance/resistance instead of reacting to it with hostility or defensiveness.

Be realistic and flexible.   Remember that there are limits to what a helper can do. Know your own personal and professional limits. If your expectations for growth, development, and change exceed the client's, you can end up in an adversarial relationship. Rigid expectations of the client and yourself become self-defeating.

Establish a "just society" with your client.   Deal with the client's feelings of coercion. Provide what Smaby and Tamminen (1979) called a "two-person just society" (p. 509). A just society is based on mutual respect and shared planning. Therefore, establish as much mutuality as is consonant with helping goals. Invite participation. Help clients participate in every step of the helping process and in all the decision making. Share expectations. Discuss and get reactions to helping procedures. Explore the helping contract with your clients and get them to contribute to it.

Help clients search for incentives for moving beyond resistance.   Help clients find incentives for participating in the helping process. Use client self-interest as a way of identifying these. Use brainstorming as a way of discovering possible incentives. For instance, the realization that he or she is going to remain in charge of his or her own life may be an important incentive for a client.

Tap significant others as resources.   Do not see yourself as the only helper in your clients' lives. Engage significant others such as peers and family members in helping clients face their reluctance and resistance. For instance, lawyers who belong to Alcoholics Anonymous might be able to deal with a lawyer's reluctance to join a treatment program more effectively than you can.

Employ clients as helpers.   If possible, find ways to get reluctant and resistant clients into situations where they are helping others. The change of perspective can help clients come to terms with their own unwillingness to work. One tactic is to take the role of the client in the interview and manifest the same kind of reluctance or resistance he or she is. Have the client take the counselor role and help you overcome your unwillingness to work or cooperate. Group counseling, too, is a forum in which clients become helpers. One person who did a great deal of work for Alcoholics Anonymous had a resistant alcoholic go with him on his city rounds, which included visiting hospitals, nursing homes for alcoholics, jails, flophouses, and down-and-out people on the streets. The alcoholic saw through all the lame excuses other alcoholics offered for their plight. After a week he joined AA himself.

In summary, do not avoid dealing with reluctance and resistance, but do not reinforce these processes in clients. Work with your client's unwillingness and become inventive in finding ways of dealing with it. As outlined earlier in the brief discussion on case management, the problem-management model itself can be applied to reluctance and resistance.

STEP I-A AND ACTION

If what you do in Step I-A is worth its salt, clients will move closer to problem-managing action. There are a number of ways of looking at the relationship between this first step of the model and action. Some clients seem to need only this first step of the helping model: they spend a relatively limited amount of time with a helper, they tell their story in greater or lesser detail, and then they go off and evidently manage quite well on their own. Here are two ways in which clients may need only the first step of the helping process.

A declaration of intent and the mobilization of resources.   For some clients the very fact that they approach someone for help may be sufficient to help them begin to pull together the resources needed to manage their problem situations more effectively. For these clients, going to a helper is a declaration, not of helplessness, but of intent: "I'm going to do something about this problem situation."

> Sean was a young man from Ireland living illegally in New York. Despite the fact that he had a community of fellow countrymen to provide him support, he felt, as he put it, "hunted." He lived with the fear that something terrible was going to happen. He talked to a priest about his concerns. The priest listened carefully, and gave no advice. But the talk provided the stimulus Sean needed. He acted. He returned to Ireland, became skilled in computers, moved to Britain, and got a job with the English arm of a German computer firm. Just as important, he felt "whole" again.

Merely seeking help can trigger a resource-mobilization process. Once they begin to mobilize their resources, they begin to manage their lives quite well on their own and no longer need the services of a helper.

Coming out from under self-defeating emotions.   Some clients come to helpers because they are incapacitated, to a greater or lesser degree, by feelings and emotions. Often, when helpers show such clients respect, listen carefully to them, and understand them in a nonjudgmental way, their self-defeating feelings and emotions subside. Once this happens, they are able to call on their own inner and environmental resources and begin to manage the problem situation that precipitated the incapacitating feelings and emotions. In short, they move to action. These clients, too, seem to be "cured" merely by telling their story. Once they get out from under the emotions that have been burdening them, they once more take charge of their lives.

> Katrina was very depressed after undergoing a hysterectomy. She had only two relatively brief sessions with a counselor. Once helped to sort through the emotions she was feeling, Katrina discovered that her predominant emotion was *shame*. She felt wounded and exposed *to herself* and guilty about her incompleteness. Once she saw what was going on she was able to pick up her life once more.

Such clients may even say something like "I feel relieved even though I haven't solved my problem." Often they feel more than relieved; they feel empowered by their interactions, however brief, with a caring helper.

It goes without saying that not all clients fall into these two categories. Those who do not move readily to action need further help. Skilled helpers, in collaboration with their clients, are able to help clients discover what further services they need beyond being helped to tell their stories.

ONGOING EVALUATION

The "E" in the center of each of the stages of the helping process illustrated in Figure 7-1 reminds us that evaluation should be built into the entire model. Evaluation is client-centered. It means making sure that the helping process is continually benefitting the client. Evaluation promotes friendly controls, not rigid attempts to make sure that everything goes by the book. Evaluation in the best sense means making sure that helping remains a learning process in which counselor and client together learn about themselves and about the process of change.

Appraisal that comes at the end of a process is often too late. How does a client benefit when, at the end of a series of helping sessions, we say it didn't work? Ongoing evaluation is much more positive. It helps both client and helper learn from what they have been doing, celebrate what has been going well, and correct what has been going poorly. Therefore, a series of evaluation questions will be listed for each of the nine steps of the helping model. In the beginning, you might well ask yourself these questions—at least the relevant ones—explicitly. If you do this, you will come to a point where you do not have to ask the questions explicitly because you will have developed an evaluative sense. That is, some part of you will always be monitoring your interactions with your clients in an unobtrusive and constructive way. It is this sense that will have you at times stop the interaction with the client and say something like "Let's stop a moment and take stock of how well we're doing here."

The following page lists evaluation questions for Step I-A. These questions serve as a summary of the main points of the chapter.

EVALUATION
QUESTIONS FOR STEP I-A

How effectively am I doing the following?

Telling the Story

- Using a mix of attending, listening, empathy, and probing to help clients tell their stories in terms of specific experiences, behaviors, and feelings
- Using probes when clients get stuck
- Understanding blocks to client self-disclosure and providing a mix of support and challenge for clients having difficulty talking about themselves

Establishing a Working Alliance

- Developing a collaborative working relationship with the client
- Using the relationship as a vehicle for social-emotional reeducation
- Not doing for clients what they can do for themselves

Ongoing Assessment

- Using assessment as an instrument of mutual learning
- Noting client resources, especially unused resources
- Listening to clients through the larger contexts of their lives
- Getting an initial feeling for the severity of a client's problems and his or her ability to handle them

Managing Client Reluctance and Resistance

- Understanding my own reluctance to grow and not being surprised by client reluctance
- Accepting client reluctance and resistance and working with it rather than against it
- Seeing reluctance and resistance and other difficulties in the helping process as problems in their own right and using the problem-management approach to address them

Evaluation

- Keeping an evaluative eye on the entire process with the goal of adding value through each interaction and of making each session better

CHAPTER 8

STEP I-B: HELPING CLIENTS CHALLENGE THEMSELVES

Introduction to Challenging

We have noted that effective helpers not only are committed to understanding clients, including the ways in which they experience themselves and the world, but also are reality testers. Contrary to one version of the client-centered credo, empathic understanding is often not enough. Senge (1990) talked about two kinds of learning. Adaptive learning is about coping. Generative learning, on the other hand, is about creating and requires new ways of looking at one's inner and outer worlds. For this, clients need to be invited to challenge themselves to change. Indeed, writers who have emphasized helping as a social-influence process have always seen some form of challenge as central to helping (Bernard, 1991; Dorn, 1984, 1986; Ellis, 1987a, 1987b; Ellis & Dryden, 1987; Strong & Claiborn, 1982). If, like most of us, clients use only a fraction of their potential in managing their problem situations and developing unused resources, then challenge has a place in helping.

Put most simply, challenge is an invitation to examine internal (cognitive) or external behavior that seems to be self-defeating, harmful to others, or both—and to change that behavior. Note that the word *challenge* is used here rather than the more biting term *confrontation*. Many see confronting and being confronted as unpleasant experiences. The history of psychology has seen periods when irresponsible confrontation was justified as "honesty" (Egan, 1970). In contrast, see how effective challenge is with Alicia, a member of a counseling group who experiences herself as unattractive.

GROUP COUNSELOR: You say that you're unattractive, and yet you have talked about how you get asked out a lot. I don't find you unattractive myself. And, if I'm not mistaken, I see people here react to you in ways that say they like you. Help me pull all this together.

ALICIA: What you say is probably true, and it helps me clarify what I mean. First of all, I'm no raving beauty, and when others find me attractive, I think that means that they find me intellectually interesting, a caring person, and things like that. At times I wish I were more physically attractive, though I feel ashamed when I say things like that. The real issue is that much of the time I *feel* unattractive. And sometimes I feel most unattractive at the very moment people are telling me directly or indirectly that they find me attractive.

GROUP MEMBER: So you've gotten into the habit of telling yourself in various ways that you're unattractive. It sounds like a bad habit.

ALICIA: It's a lousy habit. If I look at my early home life and my experiences in grammar school and high school, I could probably give you the long, sad story of how it happened. But that was then and this is now.

Since the counselor's experience of Alicia is so different from her experience of herself, he invites her to explore the difference in the group. Her self-exploration clarifies the issue greatly. It might not yet be clear to her how she is to get rid of this "lousy habit," but the issue is on the table.

I. Current scenario   II. Preferred scenario   III. Strategy: Getting there

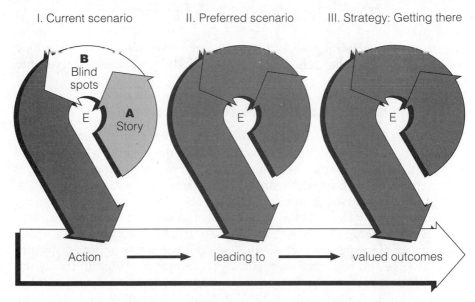

FIGURE 8-1:
Step I-B: Identifying and Challenging Blind Spots

Figure 8-1 adds this second step to Stage I. However, while challenge is discussed here as a separate step, in reality it involves a set of communication skills that are to be used in all stages and steps of the helping process. Clients can be helped to challenge themselves to:

- talk about their problems when they are reluctant to do so;
- clarify problem situations in terms of specific experiences, behaviors, and feelings when they are being vague or evasive;
- develop new perspectives on themselves, others, and the world when they prefer to cling to distortions;
- say what they want, review alternative scenarios, critique them, develop goals, and commit themselves to reasonable agendas instead of wallowing in the past;
- search for ways of getting what they want even in the face of obstacles;
- spell out specific plans instead of taking a scattered, hit-or-miss approach to change;
- persevere in the implementation of these plans when they are tempted to give up.

In a word, counselors can help clients challenge themselves to engage more effectively in all the stages and steps of problem management and constructive change. In this chapter the nature of challenging is discussed. Chapter 9 focuses on the communication skills and strategies involved in challenging.

THE GOALS OF CHALLENGING

There are two general goals of this step—one cognitive, developing new perspectives; the other behavioral, translating these new perspectives into action.

Developing New Perspectives

If challenge is successful, clients are helped to replace blind spots with new perspectives. If counselors see the world only as their clients see it, they will have little to offer them. The client's implicit interpretation of the world is often distorted, at least with respect to problems or unused opportunities. Effective helpers assume, however, that clients have the resources to see the world, especially the world of their own behavior, in a less distorted way and to act on what they see. Since blind spots and distortions are so common in human experience, challenge, especially in terms of promoting attitude change and cognitive restructuring, is central to counseling and therapy. Skilled counselors help clients move from what the Alcoholics Anonymous movement calls "stinkin' thinkin'" to "healthy thinking" (Kendall, 1992).

There are many names for this process: seeing things more clearly, getting the picture, getting insights, developing new perspectives, spelling out implications, transforming perceptions, developing new frames of reference, looking for meaning, shifting perceptions, seeing the bigger picture, developing different angles, seeing things in context, context-breaking, rethinking, getting a more objective view, interpreting, overcoming blind spots, second-level learning, double-loop learning (Argyris, 1982), thinking creatively, reconceptualizing, discovery, having an "a-ha" experience, developing a new outlook, questioning assumptions, getting rid of distortions, relabeling, and making connections. Add to this such terms as framebreaking, framebending, and reframing. All of these imply some kind of cognitive restructuring, developing understanding or awareness that is needed or useful for engaging in problem-managing action. Research has shown that the development of new perspectives is highly prized by clients (Elliott, 1985).

Linking New Perspectives to Action

Challenge is successful to the degree that it helps clients translate new perspectives into constructive action. For instance, Sally, a battered woman who has finally left her spouse, is helped to see how she continues to play the victim game in her relationships with other men. Once she realizes this, she changes to more adult-adult interactions with men. Kent, a 23-year-old who is HIV-positive, knows that he must change his sexual practices but finally does so only when challenged by a caring internist. Ned, a man who thought that his was a contemporary, non-sexist marriage, comes to realize that he still sees such things as household chores as "women's work." Once he gets over the shock of hearing within himself an oink or two of the male chauvinist pig, he must translate his discovery into action—that is, a more equitable sharing of household and child-rearing chores.

A caution is in order. Overstressing insight and self-understanding can actually stand in the way of action instead of paving the way for it. The search for insight can too easily become a goal in itself. Unfortunately, in psychology—especially the pop version—much more attention has been devoted to developing insight than to linking insight to action. The former is often presented as sexy; the latter is work. I do not mean to imply that achieving useful insights into oneself and one's world is not hard work and often painful. Ned's discovery was painful. But if the pain is to be turned into gain, behavioral change is required—whether the behavior is internal, external, or both. Ned had to work at changing his attitudes, keeping them changed, and delivering on the home front.

WHAT NEEDS TO BE CHALLENGED?

You will soon learn from experience the areas in which clients need to challenge themselves. There is no one formula for everyone. Here are some common areas, each of which will be illustrated with examples:

- Failure to own problems
- Failure to define problems in solvable terms
- Faulty interpretations of critical experiences, behaviors, and feelings
- Evasions, distortions, and game playing
- Failure to identify or understand the consequences of behavior
- Hesitancy or unwillingness to act on new perspectives.

This list is not exhaustive, but it's a good start. We human beings are very inventive when it comes to self-deception and avoidance.

Challenge Clients to Own Their Problems and Opportunities

It is all too common for clients to refuse to take responsibility for their problems and lost opportunities. Instead, there is a whole list of outside forces and other people who are to blame. Therefore, clients need to challenge themselves to own the problem situation. Here is the experience of one counselor who had responsibility for about 150 young men in a youth prison within the confines of a larger central prison.

> I believe I interviewed each of the inmates. And I did learn from them. What I learned had little resemblance to what I had found when I read their files containing personal histories, including the description of their crimes. What I learned when I talked with them was that they didn't belong there. With almost universal consistency they reported a "reason" for their incarceration that had little to do with their own behavior. An inmate explained with perfect

sincerity that he was there because the judge who sentenced him had also previously sentenced his brother, who looked very similar; the moment this inmate walked into the courtroom he knew he would be sentenced. Another explained with equal sincerity that he was in prison because his court-appointed lawyer wasn't really interested in helping him. (L. M. Miller, 1984, pp. 67-68)

It is easy to shake our heads when others do this. But we have only to review our own histories to realize that we do the same thing. This is not to deny that we can be victimized by others at times. It simply is not easy for people to accept responsibility for their acts and the consequences of their acts.

Not only problems but also opportunities need to be owned by clients. As Wheeler and Janis (1980) noted, "Opportunities usually do not knock very loudly, and missing a golden opportunity can be just as unfortunate as missing a red-alert warning" (p. 18). An out-of-work social worker who had been out of work for over six months once told me that he had heard that the city was hiring social workers for a new "help the neighborhoods help themselves" program. When I asked him whether he had gotten one of these positions, he said he had decided that it was probably only a rumor. Eventually, he discovered that the rumor was fact and set up an interview. But his first impulse was not to own — let alone seize — the opportunity.

Carkhuff (1987) talked about owning problem situations and unused opportunities in terms of personalizing. Suppose a client feels that her business partner has been pulling a fast one on her. Consider the difference between the following helper statements:

STATEMENT A: You feel angry because he unilaterally made the decision to close the deal on his terms.
STATEMENT B: You're angry because your interests were ignored.
STATEMENT C: You're furious because you were ignored, your interests were not taken into consideration, maybe you were even financially victimized, and you let him get away with it.

These three statements become progressively more personal. Personalizing means helping clients understand that in some situations, they have some responsibility for creating or at least perpetuating their problem situations. Statement C does precisely that. If clients are to manage problem situations, they must own them.

Challenge Clients to State Their Problems as Solvable

Haley (1976) said that if "therapy is to end properly, it must begin properly — by negotiating a solvable problem" (p. 9). It is not uncommon for clients to state problems so that they seem unsolvable. This justifies a "poor me" attitude and a failure to act. It may also elicit the sympathy of the counselor.

UNSOLVABLE PROBLEM: In sum, my life is miserable now because of my past. My parents were indifferent to me and at times even unjustly hostile.

> If only they had been more loving, I wouldn't be in this mess. I am the failed product of an unhappy environment.

Of course, clients will not use this rather stilted language, but the message is common enough. The point is that the past cannot be changed. Clients can change their attitudes about the past and deal with the present consequences of the past. Therefore, when a client defines the problem exclusively in terms of the past, the problem cannot be solved. "You certainly had it rough in the past and are still suffering from the consequences now" might be the kind of response that such a statement is designed to elicit. The client needs to move beyond such a point of view.

SOLVABLE PROBLEM: Over the years I've been blaming my parents for my misery. I still spend a great deal of time feeling sorry for myself. As a result, I sit around and do nothing. I don't make friends, I don't involve myself in the community, I don't take any constructive steps to get a decent job.

This message is quite different from that of the previous client. Stated in this way, the problem can be managed. The client can stop wasting her time blaming her parents, since she cannot change them; she can increase her self-esteem through constructive action and therefore stop feeling sorry for herself; and she can develop the interpersonal skills and courage she needs to enter more effectively into relationships with others. Effective helpers invite clients to state problems as solvable.

CLIENT: I can't get her out of my mind. And I can't start thinking of going out with other women.
HELPER A: It's devastating. You feel it's almost too much to cope with.
HELPER B: The misery of losing her just won't go away. . . . You've used the phrase "I can't" a couple of times. Could you tell me how this "I can't" feels for you?

The first helper responds with empathy alone. The second helper is empathic, but then gently invites the client to explore the "I can't."
    Even when a client one way or another says "I don't do this or that," it is not clear what this means. It can mean, "I don't want to" or "I don't know how" or "I'm being blocked" or "I don't have the courage to." Helping the client find out which applies is important. Helping clients explore their "I don'ts" is often as useful as helping them explore their "I can'ts." Helpers must assist clients to cast problems in solvable terms. Otherwise, problems will become plights. Problems can be managed; plights can only be endured.

Challenge Clients to Move Beyond Faulty Interpretations

The social sciences have a way of discovering things, highlighting them for a while, letting them recede into the background, and then discovering them once more. Under the rubric "constructivism" (*see* Borgen, 1992, pp. 120–121; Mahoney, 1991; Mahoney & Patterson, 1992, pp. 671–674; Neimeyer, 1993), we have rediscovered that human beings are not entirely rational and that mental images or cognitions do not always correspond to reality. According to the constructivist approach, reality as perceived by individuals is, to a degree, a construction — we actively construct the realities to which we respond. Cognitions are not always based on reality. This is part of the shadow side of human thinking. Of course, if each human being were totally constructivist in his or her approach to reality, this would lead to chaos and eliminate the ability to communicate: "We're all mad here." On the other hand, knowing that everyone, including our clients, constructs reality to a degree and that these constructions can contribute to their problems in living is very useful. Challenge means, in part, helping clients explore their constructions and the actions that follow from them and then, using their ability to construct reality, helping them *reconstruct* their views of themselves and their worlds in more self-enhancing ways.

Often enough clients fail to manage problem situations and develop opportunities because of the ways in which they interpret, understand, or label their experiences, behaviors, and feelings. For instance, Lila, a competent person, may be unaware of her competence; see herself as incompetent; or label her competence as "pushiness" when it is, in fact, assertiveness. Or Dudley, an aggressive and sometimes even violent person, may be unaware that he is aggressive; see himself as merely asserting his rights; or label the assertive actions of others as forms of aggression. Both need to move beyond these faulty interpretations. Here are a number of examples of helpers' inviting clients to challenge their interpretations.

The first client, Dwayne, is a single, middle-aged man who is involved in an affair with a married woman. The helper is getting the feeling that the client is being used but does not realize it.

HELPER: How often does she call you?
DWAYNE: She doesn't. I always contact her. That's the way we've arranged it.
HELPER: I'm curious why you've arranged it that way.
DWAYNE: Well, she wanted it that way.
HELPER: In mutual relationships, people are eager to get in touch with each other. I don't sense that in your relationship, but I could be wrong.
DWAYNE: Well, it's true that in some ways I'm more eager — maybe more dependent.

The helper is inviting the client to explore the relationship more fully without telling the client that there is something wrong with it.

In the following example, the client, Leslie, an 83-year-old woman, is a resident of a nursing home. She is talking to one of the nurses.

LESLIE: I've become so lazy and self-centered. I can sit around for hours and just reminisce . . . letting myself think of all the good things of the past — you know, the old country and all that. Sometimes a whole morning can go by.
NURSE: I'm not sure what's so self-indulgent about that.

LESLIE: Well, it's in the past and all about myself. . . . I don't know if it's right.
NURSE: When you talk to me about it, the reminiscing sounds almost like a
    kind of meditation for you.
LESLIE: You mean like a prayer?
NURSE: Well, yes . . . like a prayer.

The client had been interpreting her internal behavior in a self-defeating way. The nurse helps her develop a new, more positive perspective.

There has been some speculation about just what makes interpretation effective. One possibility is that interpretation is effective only when it hits the mark, when it is correct. The client is moved by the truth of the challenge. A second possibility is that, within reason, the content of the interpretation is not that important. What is important is that the helper is using interpretation as an instrument to express concern, care, and involvement. Clients are moved to reconsider their experiences, behavior, and emotions because the helper cares. A third possibility is that the content of the interpretation is not as important as the fact that the very act of interpreting stimulates a self-challenging process, opening up a new avenue to the client. A fourth possibility is that interpretations provide frameworks that enable clients to better understand their experiences, behaviors, and emotions. And it could well be a combination of all of these.

Challenge Evasions and Distortions

The focus here is on the discrepancies, distortions, evasions, games, tricks, excuse making, and smoke screens that keep clients mired in their problem situations. All of us have ways of defending ourselves from ourselves, others, and the world. But our defenses are two-edged swords. Daydreaming may help me cope with the dreariness of my everyday life, but it may also keep me from doing something to better myself. Blaming others for my misfortunes helps me save face, but it disrupts interpersonal relationships and prevents me from developing a healthy sense of self-responsibility.

The purpose of helping clients challenge themselves is not to strip clients of their defenses, which in some cases could be dangerous, but to help them overcome blind spots and develop new perspectives. There is a fair degree of overlap among the categories discussed next.

Challenging discrepancies.   It can be helpful to zero in on discrepancies between what clients think or feel and what they say, between what they say and what they do, between their views of themselves and the views that others have of them, between what they are and what they wish to be, between their expressed values and their actual behavior. For instance, a helper might challenge the following discrepancies that take place outside the counseling sessions:

- Tom sees himself as witty; his friends see him as biting.
- Minerva says that physical fitness is important, but she overeats and underexercises.

- George says he loves his wife and family, but he is seeing another woman and stays away from home a lot.
- Clarissa, unemployed for several months, wants a job, but she doesn't want to participate in a retraining program.

Let's see how this kind of discrepancy can be challenged.

COUNSELOR: I thought that the retraining program would be just the kind of thing you've been looking for.
CLARISSA: Well . . . I don't know if it's the kind of thing I'd like to do. . . . The work would be so different from my last job. . . . And it's a long program.
COUNSELOR: So you feel the fit isn't good.
CLARISSA: Yeah.
COUNSELOR: Clarissa, let's switch roles. Now you're me and I'm Clarissa. Ask me a challenging question.
CLARISSA: Hmm . . . Well, Clarissa, where's the old fire to get a job? I'm not even sure you believe what you're saying about the retraining program.
COUNSELOR: That's a great question. I'd love to hear the answer, so let's switch back.
CLARISSA: I guess I've gotten lazy. . . . I don't like being out of work, but I've gotten used to it.

The counselor sees a discrepancy between what Clarissa is saying and what she is doing. She is actually letting herself slip into the culture of unemployment. The counselor finds a way to help Clarissa challenge herself. Experience teaches helpers to spot discrepancies, distortions, and games. The trick is to help clients challenge these without arousing resistance.

Challenging distortions.   Some clients cannot face the world as it is and therefore distort it in various ways. For instance,

- Arnie is afraid of his supervisor and therefore sees her as aloof, whereas in reality she is a caring person.
- Edna sees her counselor in some kind of divine role and therefore makes unwarranted demands on him.
- Nancy sees her stubbornness as commitment.
- Eric, a young gay male who is very fearful about contracting AIDS, blames his problems on an older brother who seduced him during the early years of high school.

The distortion with Eric is that he now blames his brother for all of his woes, even those he has created himself. How might a counselor help Eric challenge this view?

COUNSELOR: Eric, every time we begin to talk about your sexual behavior or almost any other problem, you bring up your brother.
ERIC: That's where it all began!
COUNSELOR: Your brother's not around any more. Tell me what Eric wants. Tell me straight.

ERIC: I want people to leave me alone.
COUNSELOR: It's hard for me to believe that because I don't think you believe it. . . . Say what you want.
ERIC: I want what everyone wants—some one person to care about me. But . . . I'm never going to have what I want.

The counselor bluntly but caringly invites Eric to let go of the past and to let himself think more clearly about the present and the future. Eric's response gets him out of the past and resets the system, as it were.

Challenging self-defeating internal dialogue.   Many clients develop ways of thinking that keep them locked into their problem situations. In his rational-emotive approach to therapy, Ellis (1987a, 1987b) has hammered away at the need to challenge the assumptions, beliefs, and self-talk that sustains clients' self-defeating patterns behavior (*see also* Bernard & DiGiuseppe, 1989; Ellis & Dryden, 1987; Huber & Baruth, 1989). He has developed a system for helping clients come to grips with what can be called self-limiting or self-defeating internal dialogue.

CLIENT: I've decided not to apply for that job.
COUNSELOR: Tell a little bit more about that.
CLIENT: Well, it's not exactly what I want.
COUNSELOR: That's quite a change from last week. It sounded then as if it was just what you wanted.
CLIENT: Well, I've thought it over. (Pauses.)
COUNSELOR: I've got a hunch based on what we've learned about your style. You have to tell me whether there's any truth to it, though. I think you've been saying something like this to yourself: "I like the job, but I don't think I'm good enough for it. If I try it, I might fall flat on my face, and that would be awful. So I'll stick to what I've got, even though I don't like it very much." Any truth in any of that?
CLIENT (pausing):   Maybe more than I'd like to admit.

Helping clients challenge their self-limiting ways of thinking is one of the most powerful methodologies for behavioral change at your disposal. And, just as negative mind-sets stand in the way of managing problem situations and developing unused opportunities, so positive mind-sets can contribute greatly to dealing creatively with them.

Some of the common beliefs that Ellis believes get in the way of effective living are these:

- *Being liked and loved*. I must always be loved and approved by the significant people in my life.
- *Being competent*. I must always, in all situations, demonstrate competence, and I must be both talented and competent in some important area of life.
- *Having one's own way*. I must have my way, and my plans must always work out.

- *Being hurt*. People who do anything wrong, especially those who harm me, are evil and should be blamed and punished.
- *Being in danger*. If anything or any situation is dangerous in any way, I must be anxious and upset about it.
- *Being problemless*. Things should not go wrong in life, and if by chance they do, there should be quick and easy solutions.
- *Being a victim*. Other people and outside forces are responsible for any misery I experience.
- *Avoiding*. It is easier to avoid facing life's difficulties than to develop self-discipline; making demands of myself should not be necessary.
- *Tyranny of the past*. What I did in the past — and especially what happened to me in the past — determines how I act and feel today.
- *Passivity*. I can be happy by avoiding, by being passive, by being uncommitted, and by just enjoying myself.

Ellis suggests that if any of these beliefs are violated in a person's life, he or she tends to see the experience as terrible, awful, even catastrophic. But catastrophizing gets clients nowhere.

As important as his work is, Ellis has his critics. First, it is too confining to base an entire approach to helping on challenging irrational beliefs. Second, there is also the danger that the helper will try, overtly or covertly, to substitute his or her values for the client's. Finally, other helpers have taken a different approach to irrational beliefs. As we have seen, Driscoll (1984) has pointed out that helpers can show clients, not that they are irrational, but that they make sense. Instead of forcing clients to see how irrationally they are thinking and acting, helpers can challenge them to find the logic embedded even in seemingly stupid ideas and behaviors. Then clients can use this logic as a resource to manage problem situations instead of perpetuating them. A psychiatrist friend of mine helped a client see the beauty, as it were, of a very carefully-constructed self-defense system. The client, through a series of mental gymnastics and external behaviors, was cocooning himself from real life. My friend helped the client see how inventive he had been and how powerful the system that he had created was. He went on to help him redirect that power.

Helpers can challenge clients to substitute self-defeating internal images and dialogues with success-oriented images and dialogues. Consider the following case.

> Robert, a computer specialist who had recently been operated on for testicular cancer, was still suffering from postoperative anxiety and depression a year after the operation. One testicle had been removed, but there was no physical reason why this should have interfered with a normal sexual life. His physical recovery had been excellent, and so his bouts of impotence were seen as psychological in origin. The counselor had Robert imagine a future in which the problems he was experiencing had disappeared. Robert, relying on his past experience, described in great detail what it was like feeling at peace rather than anxious, enthusiastic rather than depressed. He pictured himself having

normal sexual relations that were as satisfying as any he had in the past. He pictured himself feeling good about himself and carrying on a positive internal dialogue instead of putting himself down. He developed these detailed images of a better future during the counseling sessions and outside of them. These positive images gradually drove out the cognitive load with which he had become burdened.

Encouraging images of a better future constituted the centerpiece of the helping process.

Challenging games, tricks, and smoke screens.   If clients are comfortable with their delusions and profit by them, they will obviously try to keep them. If they are rewarded for playing games, inside the counseling sessions or outside, they will continue a game approach to life (*see* Berne, 1964). For instance, Clarence plays the "Yes, but . . ." game: he presents himself as one in need of help and then proceeds to show his helper how ineffective she is. Dora makes herself appear helpless and needy when she is with her friends, but when they come to her aid, she is angry with them for treating her like a child. Kevin seduces others in one way or another and then becomes indignant when they accept his implied invitations. The number of games we can play in order to avoid the work involved in squarely facing the tasks of life is endless. Clients who are fearful of changing will attempt to set off smoke screens in order to hide from the helper the ways in which they fail to face up to life. Such clients use communication in order not to communicate (*see* Beier & Young, 1984). It helps to establish an atmosphere that discourages clients from playing games. An attitude of "Nonsense is challenged here" should gently pervade the helping sessions.

Some clients play the victim game. A therapist described one of his clients like this:

> He is enamored of victimhood. A victim of parental sexual abuse himself, he now sees himself as a victim of his other members of his family and of his church. Waiters, even though they give poor service, must be overtipped because they are victims of the workplace. Workers are victims of the economy and therefore deserve more and more entitlement programs. He sees his friends as fragile, victimized themselves in a variety of ways and therefore needing his special attention. All of this is self-perpetuating and self-defeating. Helping him challenge these assumptions is very difficult because it is quite clear that he is ready to see himself victimized by the helping profession, too. There is hope, of course, because all of this victimhood has become very painful for him.

Some clients present themselves as victims of circumstances, of other people, of life itself. The victim game can be quite rewarding. Victims are noble, good, the focus of attention, and not responsible for the evil that befalls them. In the following example, Roberto has assumed the victim role in his marriage. His wife, Maria, has broken through a number of cultural taboos. She has put herself through college, gotten a job, developed it into a career, and assumed the role of mother-breadwinner. She makes more money than Roberto. His victimhood includes losing face in the community, feeling belittled by his wife's success, and being forced into a "democratic" marriage. He has created a series of minor crises at home in hopes that his wife will rethink what she is doing. She is not present for this session. Roberto has been emphasizing his woes.

HELPER: Roberto, you sound as if you are ready to give up.
ROBERTO: Do I sound that bad? You know what makes it all worse is the fact that we've got, at least financially, what most of our friends would sell their souls for.
HELPER: Let's talk off the record—just between you and me.
ROBERTO: This sounds like it's going to hurt. What are we going to talk about?
HELPER: Maybe about sabotage. Not vicious or planned, but still quite effective. . . . You've described a number of crises at home, like your son's failure to get into the preschool you wanted. Off the record, what really happened there?
ROBERTO: If Maria had been around more . . .
HELPER (leaning forward and in a gentle voice):   Come on. We're off the record.
ROBERTO: Well, so I didn't do everything I could. So what? I wasn't exactly in the mood to.
HELPER (gently):   Roberto, we're still off the record.

The helper uses an "off the record" approach to let Roberto know that nothing in this session will be shared by him with Maria. Even then Roberto becomes mildly defensive and resistant. They go on to talk about the ways in which Roberto has been using benign forms of sabotage to get what he wants. In later sessions Maria and Roberto struggle with merging the Latino and the Anglo cultures in their marriage.

Challenging excuses.   Snyder, Higgins, and Stucky (1983; *see also* Halleck, 1988; Snyder, 1984; Snyder & Higgins, 1988) have examined excuse-making behavior in depth. Excuse making, of course, is universal, part of the fabric of everyday life. Like games and distortions, it has its positive uses in life. Even if it were possible, there is no real reason for setting up a world without myths. On the other hand, excuse making, together with avoidance behavior, probably contributes a great deal to the "psychopathology of the average." Roberto tells the helper that he engaged in benign sabotage because he wasn't ready for the changes in style that his wife's behavior was demanding from him. Halleck claimed that the helper's task is to help clients maximize their functioning and make use of all their capacities. He described a case in which an injured worker was excused from manual labor but was challenged when he thought he should be excused from all work.

We have only skimmed the surface of the games, evasions, tricks, distortions, excuses, rationalizations, and subterfuges resorted to by clients (together with the rest of the population). Skilled helpers are caring and empathic, but they do not let themselves be conned. That helps no one.

Challenge Clients to Explore the Consequences of Their Behavior

One way of helping clients get new and more creative perspectives on themselves and their behavior is to help them explore the consequences of patterns of behavior currently in place. Let's return to Roberto and his mild forms of sabotage. He refers to them as "delaying tactics."

HELPER: It might be helpful to see where all of this is leading.
ROBERTO: What do you mean?
HELPER: I mean let's review the impact your delaying tactics have had on Maria and your marriage. And then let's review where these tactics are most likely to lead.
ROBERTO: Well, I can tell you one thing. She's become even more stubborn.

Through their discussion Roberto discovers that his sabotage is not only not working for him but that it is actually working against him. His campaign is headed in the wrong direction. Compromise is becoming less and less likely.

The consequences of a client's behavior can be positive but unappreciated. Consider the case of a client in a rehabilitation unit after a serious accident. She has been sticking to the program, but progress is slow.

CLIENT: Sometimes I think that the courageous thing to do is just to chuck all this stuff, admit that I'm a cripple, and get on with life. I don't want to delude myself that I'm going to be a whole human being again. I'm not. And I'm not a hero.
COUNSELOR: I see the depression's at you again.
CLIENT: It's not at me—it's got me by the throat.
COUNSELOR: I wish I had a videotape.
CLIENT: Of what?
COUNSELOR: Of the way you could, or rather couldn't, move that left arm of yours six weeks ago. . . . Move it now.
CLIENT (moving her arm a bit):   You mean people can notice a difference? It's so hard seeing that the exercises make a difference.
COUNSELOR: Sure. You know the exercises are making a difference, but you hate the snail's pace.
CLIENT: By now you know that patience has never been one of my strong points.

Helping the client look at the positive consequences of her behavior, in this case the exercises, is a supportive gesture on the part of the helper. His challenge is rooted in his empathic understanding of her frustration.

Challenge Clients to Act

The ultimate goal of challenging can be stated in one word: action. New perspectives and heightened awareness are not goals in themselves. Action includes:

- starting activities related to managing problems and developing opportunities;

- continuing and increasing activities that contribute to problem management and opportunity development;
- stopping activities that either cause problems and limit opportunities or stand in the way of problem management and opportunity development.

These activities can be internal, external, or a combination of both. If Roberto is going to manage the conflict between himself and his wife better, he needs to challenge himself to:

- start thinking of his wife as an equal in the relationship, start understanding her point of view, and start imagining what an improved relationship with her might look like;
- continue exploring with her the ways he contributes to their difficulties and increase the number of times he tells himself to let her live her life as fully as he wants to live his own;
- stop telling himself that she is the one with the problem, stop seeing her as the offending party when conflicts arise, and stop telling himself there is no hope for the relationship.

That is, he has to put his internal life in order and begin to mobilize internal resources.

Similarly, if problems are to be managed and opportunities developed, then some external behaviors need to be started, some continued or increased, and some stopped. For instance, if Roberto is going to do his part in developing a better relationship with his wife, he needs to challenge himself to:

- start spending more time looking at his own career, start sharing his feelings with her, start engaging in mutual decision making, and start taking more initiative in household chores and child care;
- continue visiting her parents with her and increase the number of times he goes to business-related functions with her;
- stop criticizing her in front of others, stop creating crises at home and assigning the blame for them to her, and stop making fun of her business friends.

With the counselor's help, Roberto can challenge himself to do all of these things.

LINKING CHALLENGE TO ACTION

A counselor friend of mine makes an audiotape of each session, gives it to the client at the end of the session, and makes some suggestions on what to listen for. This makes the client do something about the process. But it can also influence the client to act in other ways. Once he gave a client the tape and said, "Listen to how you responded when I asked you about what you might do about patching up your relationship with your father. Up to that point your voice was strong. But when you talked about what you might do, your voice got much softer and more tentative. See if you can notice any difference between what you say and how you say it."

Sometimes new perspectives lead to action. At other times action generates new perspectives.

> Woody, a sophomore in college, came to the student counseling services with a variety of interpersonal and somatic complaints. He felt attracted to a number of women on campus but did very little to become involved with them. After exploring this issue briefly, he said to the counselor: "Well, I just have to go out and do it." Two months later he returned and said that his experiment had been a disaster. He had gone out with a few women, but none of them really interested him. Then he did meet someone he liked quite a bit. They went out a couple of times, but the third time he called, she said that she didn't want to see him anymore. When he asked why, she muttered vaguely about his being too preoccupied with himself and ended the conversation. He felt so miserable he returned to the counseling center. He and the counselor took another look at his social life. This time, however, he had some experiences to probe. He wanted to explore his being "too preoccupied with himself."

This student put into practice Weick's (1979) dictum that chaotic action is sometimes preferable to orderly inactivity. Some of Woody's learnings were painful, but now there is a chance of examining his interpersonal style much more concretely. Similarly, one of the assumptions of Alcoholics Anonymous is that people sometimes need to act themselves into new ways of thinking rather than think themselves into new ways of acting.

The Shadow Side of Challenging

Challenge often deals with murkier experiences, behaviors, and feelings on the part of both clients and helpers.

Shadow-Side Responses to Challenge

Even when challenge is a response to a client's plea to be helped to live more effectively, it can precipitate some degree of disorganization in the client. Different writers refer to this experience under different names: crisis, disorganization, a sense of inadequacy, disequilibrium, and beneficial uncertainty (Beier & Young, 1984). As the last of these terms implies, counseling-precipitated crises can be beneficial for the client. Whether they are or not depends, to a great extent, on the skill of the helper. Even when challenge is invitation rather than coercion, clients resist.

Even when an invitation to self-challenge or a direct challenge is accurate and delivered caringly, some still dodge and weave. Cognitive-dissonance theory (Festinger, 1957) gives us some insight into the dynamics of this.

Since dissonance (discomfort, crisis, disequilibrium) is an uncomfortable state, the client will try to get rid of it. According to dissonance theory, there are five typical ways in which people experiencing dissonance attempt to rid themselves of this discomfort. Let's examine them briefly as they apply to being challenged. Since these responses are forms of resistance, review the ways in which resistance can be avoided and managed (Chapter 7) and see how they might apply to the following examples.

1. Discredit challengers.   The challenger is confronted and discredited. Some attempt is made to point out that he or she is no better than anyone else. In the following example, the client has been discussing her marital problems and has been invited to take a second look at her behavior by the helper.

CLIENT: It's easy for you to sit there and suggest that I be more responsible in my marriage. You've never had to experience the misery in which I live. You've never experienced his brutality. You probably have one of those nice middle-class marriages.

Counterattack is a common strategy for coping with challenge. How do you think this particular instance of counterattack should be handled?

2. Persuade challengers to change their views.   In this approach, challengers are reasoned with. They are urged to see what they have said as misinterpretations and to revise their views. In the following example, the client has been talking about the way she blows up at her husband and has been invited to explore the consequences of this pattern of venting her anger.

CLIENT: I'm not so sure that my anger at home isn't called for. I think that it's a way in which I'm asserting my own identity. If I were to lie down and let others do what they want, I would become a doormat at home. And, as you have helped me see, assertiveness should be part of my style. I think you see me as a fairly reasonable person. I don't get angry here with you because there is no reason to.

Sometimes a client like this will lead an unwary counselor into an argument about the issue in question. How would you respond to this client?

3. Devalue the issue.   This is a form of rationalization. A client who is being invited to challenge himself about his sarcasm points out that he is rarely sarcastic, that "poking fun at others" is just good-natured fun, that everyone does it, that it is a very minor part of his life and not worth spending time on. The client has a right to devalue a topic if it really isn't important. The counselor has to be sensitive enough to discover which issues are important and which are not. How would you handle this client's devaluing the issue?

4. Seek support elsewhere for the views being challenged.   Some clients leave one counselor and go to another because they feel they aren't being understood. They try to find helpers who will agree with them. This is an extreme way of seeking support of one's own views elsewhere. But a client can remain with a counselor and still use this strategy by offering evidence that others contest the helper's point of view.

CLIENT: I asked my wife about my sarcasm. She said she doesn't mind it at all. And she said she thinks that my friends see it as humor and as a part of my style.

This is an indirect way of telling the counselor she is wrong. The counselor might well be wrong, but if the client's sarcasm is really dysfunctional in his interpersonal life, the counselor should find some way of pressing the issue. What would you do in this case?

5. Respond to the invitation to self-challenge but then don't act on it. The client can agree with the counselor as a way of dismissing an issue. However, the purpose of challenging is not to get the client's agreement but to develop new perspectives that lead to constructive action. Consider this client.

CLIENT: To tell you the truth, I'm pretty lazy and manipulative. And I'm not very clever.

In truth this client was lazy and manipulative and was being both in this instance. But he was very clever. It was up to the helper to break through this screen of "honesty." What would you do in this case?

Challenge and the Shadow Side of Helpers

Two shadow-side areas are addressed here — the reluctance of some helpers to challenge clients or invite them to challenge themselves and the blind spots helpers themselves have.

The MUM effect.   Initially some counselor trainees are quite reluctant to help clients challenge themselves. They become victims of what has been called the "MUM effect," the tendency to withhold bad news even when it is in the other's interest to hear it (Rosen & Tesser, 1970, 1971; Tesser & Rosen, 1972; Tesser, Rosen, & Batchelor, 1972; Tesser, Rosen, & Tesser, 1971). In ancient times, sometimes the person who bore bad news to the king was killed. This obviously led to a certain reluctance on the part of messengers to bring such news. Bad news — and, by extension, the kind of bad news that is involved in any kind of invitation to self-challenge — arouses negative feelings in the challenger, no matter how he or she thinks the receiver will react. If you are comfortable with the supportive dimensions of the helping process but uncomfortable with helping as a social-influence process, you could fall victim to the MUM effect and become less effective than you might be.

Reluctance to challenge is not a bad starting position. In my estimation, this is far better than being too eager to challenge. However, all helping, even the most client-centered, involves social influence. It is important for you to understand your reluctance (or eagerness) to challenge — that is, to challenge yourself on the issue of challenging and on the very notion of helping as a social-influence process. Some of the things that trainees discover when they examine how they feel about challenging others follow:

- "I am just not used to challenging others. My interpersonal style has had a lot of the live-and-let-live in it. I have misgivings about intruding into other people's lives."
- "If I challenge others, then I open myself to being challenged. I may be hurt, or I may find out things about myself that I would rather not know."
- "I might find out that I like challenging others and that the floodgates will open and my negative feelings about others will flow out. I have some fears that deep down I am a very angry person."
- "I am afraid that I will hurt others, damage them in some way or other. I have been hurt or I have seen others hurt by heavy-handed confrontations."
- "I am afraid that I will delve too deeply into others and find that they have problems that I cannot help them handle. The helping process will get out of hand."
- "If I challenge others, they will no longer like me. I want my clients to like me."

It is useful to examine where you stand in terms of helping process, helping values, and personal feelings with respect to inviting others to challenge themselves and therefore opening yourself up to challenge. Finally, being willing to challenge responsibly is one thing; having the skills to do so is another. Chapter 9 focuses on the skills needed to invite others to take a better look at themselves.

Helpers' blind spots.   Since helpers are as human as their clients, they too can have blind spots that detract from their ability to help. For instance, in one study (Atkinson, Worthington, Dana, & Good, 1991) counselors were almost unanimous in their preference for a feeling approach to counseling, whereas the majority of male clients preferred either a thinking or acting orientation. Helpers sometimes blame clients for lack of movement when in reality they don't know what to do to get the client moving or are afraid to do so. They can see themselves as challenging when they are, in fact, punitive. Or they don't realize that the client has taken charge of the helping process and is manipulating them. One of the critical responsibilities of supervisors is to help counselors identify their blind spots and learn from them. Once out of training, skilled helpers use different forums or methodologies to continue this process, especially with difficult cases. They ask themselves, "What am I missing here?" They take counsel with colleagues. Without becoming self-obsessed, they scrutinize and challenge themselves and the role they play in the helping relationship.

CHAPTER 9

COMMUNICATION SKILLS III: SKILLS AND GUIDELINES FOR EFFECTIVE CHALLENGING

Counselors can use a number of approaches to help clients challenge themselves constructively. This chapter discusses and illustrates the skills of information sharing, advanced empathy, helper self-disclosure, immediacy, and focused summarizing. And, since the way in which helpers go about challenging clients or inviting them to challenge themselves can make or break the challenging process, principles of effective challenging are also reviewed.

New Perspectives through Information

Sometimes clients are unable to explore their problems fully and proceed to action because they lack information of one kind or another. For instance, it helps many clients to know that they are not the first to try to cope with a particular problem. In addition, there is much to be learned from other people's experience. Accurate information can help clients accept themselves and their problem situations in a more upbeat way.

The skill or strategy of information sharing is included under challenging skills because it helps clients develop new perspectives on their problems. It includes both giving information and correcting misinformation. In some cases the information can prove to be quite confirming and supportive. For instance, a parent who feels responsible following the death of a newborn baby may experience some relief through an understanding of the features of the Sudden Infant Death Syndrome. This information does not solve the problem, but the parent's new perspective can help him or her handle self-blame.

The new perspectives clients gain from information sharing can also be quite challenging. Consider the following example.

> Troy was a college student of modest intellectual means. He made it through
> school because he worked very hard. In his senior year he learned that a number of his friends were going on to graduate school. He, too, applied to a
> number of graduate programs in psychology. He came to see a counselor in the
> student services center after being rejected by all the schools to which he applied. In the interview it soon became clear to the counselor that Troy thought
> that many, perhaps even most, college students went on to graduate school.
> After all, most of his closest friends had been accepted in one graduate school
> or another. The counselor shared with him the statistics of what could be called
> the educational pyramid—the decreasing percentage of students attending
> school at higher levels. Troy did not realize that just finishing college made him
> part of an elite group. Nor was he completely aware of the extremely competitive nature of the graduate programs in psychology to which he had applied. He
> found much of this relieving, but then found himself suddenly faced with
> what to do now that he was finishing school. Up to this point he had not

thought much about it. He felt disconcerted by the sudden need to look at the world of work.

Giving information is especially useful when lack of accurate information either is one of the principal causes of a problem situation or is making an existing problem worse.

In some medical settings, doctors team up with counselors to give clients messages that are hard to hear and to provide them with information needed to make difficult decisions. For instance, Lester, a 54-year-old accountant, has been given a series of diagnostic tests for a heart condition. The doctor and counselor sit down and talk with him about the findings. Bypass surgery is one option, but it has risks and there is no absolute assurance that the surgery will take care of all his heart problems. The counselor helps Lester cope with the news, process the information, and come to a decision.

There are some cautions to be observed in giving information. When information is challenging, or even shocking, the helper needs to be tactful and know how to help the client handle the disequilibrium that comes with the news. Also, the helper must not overwhelm the client with information. As a helper, you must make sure that the information you provide is clear and relevant to the client's problem situation. Don't let the client go away with a misunderstanding of the information. Be sure not to confuse information giving with advice giving; the latter is seldom useful. Finally, be supportive; help the client process the information. All of these cautions come into play for counselors of those who test positive for the AIDS virus.

> Angie, an unmarried woman who has just given birth to a baby boy, needs to be told that both she and her son are HIV-positive. The implications are enormous. She needs information about the virus, what she needs to do for herself and her son medically, and the implications for her sex life. Obviously the way in which this information is given to her can determine to a great extent how she will handle the immediate crisis and the ensuing lifestyle demands.

It is obvious that Angie's case is not one for an amateur. Experts, particularly those with special training in dealing with AIDS cases, know the range of ways in which clients receive the news that they have tested positive, how to provide support in each case, and how to challenge clients to rally resources and manage the crisis.

At times of major decisions, helpers should not use information as a subtle (or not too subtle) way of pushing their own values. For instance, helpers should not immediately give clients with unwanted pregnancies information about abortion clinics. Conversely, if abortion is contrary to the helper's values, then the helper needs to let the client looking for abortion counseling know that he or she cannot help her in this area.

ADVANCED EMPATHY: GETTING BEHIND CLIENTS' MESSAGES

Chapter 6 outlined the characteristics of empathy in general and of basic empathy in particular. However, as skilled helpers listen intently to clients, they often enough see clearly what clients only half see and hint at. This deeper kind of empathy involves "sensing meanings of which the client is scarcely aware" (Rogers, 1980, p. 142) or, in broader terms, the "story behind the story" (Berger, 1989). For instance, a client talks about his anger with his wife, but as he talks, the helper hears not just anger but also hurt. It may be that the client can talk with relative ease about his anger but not as easily about his feelings of hurt. In a basic empathic response, clients recognize themselves almost immediately: "Yes, that's what I meant." Since advanced empathy digs a bit deeper, clients might not immediately recognize themselves in the helper's response. That's what makes advanced empathy, while still empathy, a form of challenge.

Here are some questions helpers can ask themselves to probe a bit deeper as they listen to clients: What is this person only half saying? What is this person hinting at? What is this person saying in a confused way? What messages do I hear behind the explicit messages?

When a helper says to a client, "You feel angry because your wife has been paying a lot of attention to a younger man—angry and perhaps a bit hurt" the statement can make the client stop short because the anger is up front and the hurt, although real, is in the background. That's what makes advanced empathy challenging. Note that advanced empathic listening deals with what the client is actually saying or expressing, however confusedly, and not with the helper's interpretations of what the client is saying. Advanced empathy is not an attempt to psych the client out.

Advanced empathy focuses not just on problems, but also on unused or partially used resources. Effective helpers listen for the resources that are buried deeply in clients and often have been forgotten by them. Cummings (1979) gave an example of this kind of listening in his work with addicts.

> During the first half of the first session the therapist must listen very intently. Then, somewhere in midsession, using all the rigorous training, therapeutic acumen, and the third, fourth, fifth, and sixth ears, the therapist discerns some unresolved wish, some long-gone dream that is still residing deep in that human being, and then the therapist pulls it out and ignites the client with a desire to somehow look at that dream again. This is not easy, because if the right nerve is not touched, the therapist loses the client. (p. 1123)

Advanced empathy deals with both the overlooked positive side and the overlooked shadow side of the client's experience and behavior.

In the following example the client, a soldier doing a 5-year hitch in the army, has been talking to a chaplain about his failing to be promoted. As he talks, it becomes fairly evident that part of the problem is that he is so quiet and unassuming that it is easy for his superiors to ignore him. He is the kind of person who keeps to himself and keeps everything inside.

CLIENT: I don't know what's going on. I work hard, but I keep getting passed over when promotion time comes along. I think I work as hard as anyone else, and I work efficiently, but all of my efforts seem to go down the drain. I'm not as flashy as some others, but I'm just as substantial.

CHAPLAIN A: You feel it's quite unfair to do the kind of work that merits a promotion and still not get it.

CHAPLAIN B: It's depressing to put as much effort as those who get promoted and still get passed by. . . . Tell me more about the not-as-flashy bit. What in your style might make it easy for others not to notice you, even when you're doing a good job?

Chaplain A tries to understand the client from the client's frame of reference. He deals with the client's feelings and the experience underlying these feelings. This is basic empathy. Chaplain B, however, goes a bit further. From the context, from past interchanges, from the client's manner and tone of voice, he picks up a theme that the client states in passing: that the client is so unassuming that his best efforts go unnoticed. Advanced empathy, then, goes beyond the expressed to the partially expressed and the implied. If helpers are accurate and if their timing is good, they will assist the client to develop a new and useful perspective.

Let's take a look at the client's response to each chaplain.

CLIENT (in response to Chaplain A):   Yeah . . . I suppose there's nothing I can do but wait it out. (A long silence ensues.)

CLIENT (in response to Chaplain B):   You mean I'm so quiet I could get lost in the shuffle? Or maybe it's the guys who make more noise — my dad called them the squeaky wheels — who get noticed.

In his response to Chaplain A, the client merely retreats more into himself. However, in his response to Chaplain B, the client begins to see that his unassuming, nonassertive style may contribute to the problem situation. Once he becomes aware of the self-limiting dimensions of his style, he is in a better position to do something about it. Advanced empathy can take a number of forms. Let's consider some of them.

Helping Clients Make the Implied Explicit

The most basic form of advanced empathy is to help clients give fuller expression to what they are implying. In the following example, the client has been discussing ways of getting back in touch with his wife after a recent divorce, but when he speaks about doing so, he expresses very little enthusiasm.

CLIENT (somewhat hesitatingly):   I could wait to hear from her. But I suppose there's nothing wrong with calling her up and asking her how she's getting along.

COUNSELOR A: It seems that it would be all right for you to take the initiative to find out if everything is well with her.

CLIENT (somewhat drearily):   Yeah, I suppose I could.

COUNSELOR B: You've been talking about getting in touch with her, but, unless I'm mistaken, I don't hear a great deal of enthusiasm in your voice.

CLIENT: To be honest, I don't really want to talk to her. But I feel guilty, guilty about the divorce, guilty about seeing her out on her own. I'm taking care of her all over again. And that's one of the reasons we got divorced—I had a need to take care of her and she let me do it. I don't want to do that anymore.

Counselor A's response might have been fine at an earlier stage of the helping process, but it misses the mark here, and the client grinds to a halt. The goal of Step I-B is to help the client dig deeper. Counselor B bases her response not only on the client's immediately preceding remark but on the entire context of the story-telling process. Her response hits the mark, and the client moves forward. As with basic empathy, there is no such thing as a good advanced empathic response in itself. The question is: Does the response help the client clarify the issue more fully so that he or she might begin to see the need to act differently?

Advanced empathy is part of the social-influence process: it places demands on clients to take a deeper look at themselves. Challenges are based on a basic empathic understanding of the client and are made with genuine care and respect, but they are demands nevertheless.

Helping Clients Identify Themes

Helping clients identify and explore themes that emerge in their discussions of problem situations and unused opportunities, especially self-defeating themes, is critical. Recall the case of Roberto and Maria from Chapter 8. The counselor spends time with them together but also sees each of them alone once in a while. It is Maria's turn.

COUNSELOR: I'd like to pull a few things together. If I have heard you correctly, you currently take care of the household finances. You are usually the one who accepts or rejects social invitations, because your schedule is tighter than Roberto's. And now you're about to ask him to move because you can get a better job in Boston.

MARIA: When you put it all together like that, it sounds as if I'm running his life.

COUNSELOR: What would your reaction be if the picture were reversed?

MARIA: Hmmm . . . Well, I'd . . . hmm . . . (laughs). I'm giving myself away! I guess I wouldn't mind any one part of the picture, but I think I would resent the whole thing.

Helping Maria explore the "I am making all the big decisions for him" theme is a step forward. When she puts herself in his shoes, she doesn't like it.

Thematic material might refer to feelings (such as themes of hurt, of depression, of anxiety), to behavior (such as themes of controlling others, of avoiding intimacy, of blaming others, of overwork), to experiences (such as themes of being a victim, of being seduced, of being punished, of failing), or some combination of these. Once you see a self-defeating theme or pattern emerging from your discussions, your task is to communicate your perception to the client in a way that enables the client to check it out. Make sure that the themes you discover are based on the client's experience and are not just the artifacts of some psychological theory. Advanced empathy works because clients recognize themselves in what you say.

Helping Clients Make Connections

Clients often reveal experiences, behaviors, and emotions in a hit-and-miss way. The counselor's job, then, is to help them make the kinds of connections that provide action-oriented insights or perspectives. The following client, who has a full-time job, is finishing his final two courses for his college degree and is going to get married right after graduation. He talks about being progressively more anxious and tired in recent weeks. Later he talks about getting ready for his marriage in a few months and about deadlines for turning in papers for current courses. Still later, he talks about his need to succeed, to compete, and to meet the expectations of his parents and grandparents. He has begun to wonder whether all of this is telling him that it is a mistake to get married or that he is an inadequate person. The fact that a recent physical examination showed him to be in good health has actually made things worse.

COUNSELOR: John, there might be a simpler explanation for your growing anxiety and tiredness. One, you are really working very hard at school and at work. Two, competing as hard as you do is bound to take its physical and emotional toll. And three, the emotional drain of getting ready for a marriage is enormous for anyone, including you. I wonder how Superman might manage all this.

JOHN: I thought that the medical exam would have shown some physical basis for being so tired. The idea that I'm trying to juggle too many things never crossed my mind. Ever since I was a kid I was supposed to handle whatever came along. My grandfather did. My dad did. No complaints.

John had been talking about these four "islands" — a full-time job, finishing school, preparing for marriage, and meeting the expectations of his family — as if they were unrelated to one another. The counselor does two things. First, he helps John make the connections. Second, he provides a bit of information. Finishing school and getting ready for marriage score high on the stress scale. High expectations on the part of his family act as a multiplier. They go on to explore ways in which John can manage his stress.

HELPER SELF-DISCLOSURE

Still another skill of challenging involves the ability of helpers to constructively share some of their own experiences, behaviors, and feelings with clients (Hendrick, 1990; Mathews, 1988; Simon, 1988; Stricker & Fisher, 1990; Watkins, 1990). In one sense counselors cannot help but disclose themselves: "The counselor communicates his or her characteristics to the client in every look, movement, emotional response, and sound, as well as with every word" (Strong & Claiborn, 1982, p. 173). This is indirect disclosure. As they attend, listen, and respond, helpers should track the impressions they are making on clients.

Here, however, it is a question of direct self-disclosure. In some forms of helping, direct self-disclosure serves as a form of modeling. Self-help groups such as Alcoholics Anonymous use such modeling extensively. This helps new members get an idea of what to talk about and helps arouse the courage to do so. It is the group's way of saying "You can talk here without being judged and getting hurt." Even in one-to-one counseling dealing with alcohol and drug addiction, extensive helper self-disclosure is the norm.

> Beth is a counselor in a drug rehabilitation program. She herself was an addict for a number of years, but "kicked the habit" with the help of the agency where she is now a counselor. It is clear to all addicts in the program that the counselors there were once addicts themselves and are not only rehabilitated but intensely interested in helping others rid themselves of drugs and develop a kind of lifestyle that helps them stay drug free. Beth freely shares her experience, both of being a drug user and of her rather agonizing journey to freedom, whenever she thinks that doing so can help a client.

Ex-addicts often make excellent helpers in programs like this. They know from the inside the games addicts play. Sharing their experience is central to their style of counseling and is accepted by their clients. New perspectives are developed, and new possibilities for action are discovered. Such self-disclosure is challenging. It puts pressure on clients to talk about themselves more cogently or in a more focused way.

Helper self-disclosure is challenging for at least two reasons. First, it is a form of intimacy and, for some clients, intimacy is not easy to handle. Second, the message to the client is, indirectly, a challenging "You can do it, too," because helper revelations, even when they deal with past failures, usually deal with problem situations that have been overcome.

Research into helper self-disclosure has produced mixed and even contradictory results. Some researchers have discovered that helper self-disclosure can frighten clients or make them see helpers as less well ad-

justed. Other studies have suggested that helper self-disclosure is appreciated by clients. Some clients see self-disclosing helpers as down to earth and
honest. Since current research does not tell us a great deal, we need to stick
to common sense.

In the following example, the helper, Rick, has had a number of sessions with Tim, a client who has had a number of developmental problems
and is now taking an overly-cautious approach to almost everything in life.

RICK: In my junior year in high school I was expelled for stealing. I thought
        that it was the end of the world. My Catholic family took it as the ultimate
        disgrace. We even moved to a different neighborhood in the city.
TIM: What did it do to you?

Rick briefly tells his story, a story that includes setbacks — not unlike
Tim's — but one that eventually has a successful outcome. Rick does not
overdramatize his story. In fact, his story makes it clear that developmental
crises are normal. How they are interpreted and managed is the critical
issue.

The conflicting results of research studies tell us that, currently, helper
self-disclosure is not a science, but an art. Some guidelines for using it are
described here.

Include helper self-disclosure in the contract. In self-help groups and in the
counseling of addicts by ex-addicts, helper self-disclosure is an explicit part
of the contract. If you don't want your disclosures to surprise your clients,
let them know that you may self-disclose. Therefore, a helper might say
somewhere toward the beginning of the counseling process something like
this: "From time to time I might share with you some of my own life experiences if they can help us move forward."

Make sure that your disclosures are appropriate. Sharing yourself is appropriate if it helps clients achieve the treatment goals outlined in this helping
process. Don't disclose more than you have to. Helper self-disclosure that is
exhibitionistic or engaged in for effect is obviously inappropriate. Timing is
critical. Premature self-disclosure by a helper can turn clients off.

Keep your disclosure selective and focused. Don't distract clients with ram-
bling stories about yourself. In the following example, the helper is talking
to a first-year grad student in a clinical psychology program. The client is
discouraged and depressed by the amount of work she has to do.

COUNSELOR: Listening to you takes me right back to my own days in gradu-
        ate school. I don't think that I was ever busier in my life. I also believe that
        the most depressing moments of my life took place then. On any number of
        occasions I wanted to throw in the towel. For instance, I remember once to-
        ward the end of my third year when . . .

It may be that selective bits of this counselor's experience in graduate school
might be useful in helping the student get a better conceptual and emotional

grasp of her problems, but the counselor has wandered off into the kind of reminiscing that meets his needs rather than the client's.

Helper self-disclosure is inappropriate if it is too frequent. Some research (Murphy & Strong, 1972) has suggested that if helpers disclose themselves too frequently, clients tend to see them as phony and suspect that they have ulterior motives.

Do not burden the client. Do not burden an already overburdened client. One counselor thought that he would help make a client who was sharing some sexual problems more comfortable by sharing some of his own experiences. After all, he saw his sexual development as not too different from the client's. However, the client reacted by saying: "Hey, don't tell me your problems. I'm having a hard enough time dealing with my own." This novice counselor shared too much of himself too soon. He was caught up in his own willingness to disclose rather than its potential usefulness to the client.

Remain flexible. Take each client separately. Adapt your disclosures to differences in clients and situations. Rick shared his high-school experience with Tim relatively early on in their talks as a way of challenging Tim to stop seeing himself as unique. When asked directly, clients say that they want helpers to disclose themselves (*see* Hendrick, 1990), but this does not mean that every client in every situation wants it or would benefit from it. Self-disclosure on the part of helpers should be a natural part of the helping process, not a gambit.

IMMEDIACY: DIRECT, MUTUAL TALK

Many, if not most, clients who seek help have trouble with interpersonal relationships. This is either their central concern or part of a wider problem situation. Some of the difficulties clients have in their day-to-day relationships are also reflected in their relationships with helpers. For instance, if they are compliant outside, they are often compliant in the helping process. If they become aggressive and angry with authority figures outside, they often do the same with helpers. Therefore, the client's interpersonal style can be examined, at least in part, through an examination of his or her relationship with the helper. If counseling takes place in a group, then the opportunity is even greater. The package of skills enabling helpers to explore their relationship with their clients has been called "immediacy" by Carkhuff (*see* Carkhuff 1969a, 1969b; Carkhuff & Anthony, 1979).

Types of Immediacy in Helping

Three kinds of immediacy are reviewed here:

1. Immediacy that focuses on the overall relationship—"How are you and I doing?"

2.   Immediacy that focuses on some particular event in a session —
     "What's going on between you and me right now?"
3.   Self-involving statements.

General relationship immediacy.   General relationship immediacy refers
to your ability to discuss with a client where you stand in your overall relationship with him or her. The focus is not on a particular incident but on the
way the relationship itself has developed and how it is helping or standing
in the way of progress. In the following example, the helper is a 44-year-old
woman working as a counselor for a large company. She is talking to a 36-
year-old man she has been seeing once every other week for about two
months. One of his principal problems is his relationship to his supervisor,
who is also a woman.

COUNSELOR: We seem to have developed a good relationship here. I feel we
     respect each other. I have been able to make demands on you, and you have
     made demands on me. There has been a great deal of give-and-take in our
     relationship. You've gotten angry with me, and I've gotten impatient with
     you at times, but we've worked it out. I'm wondering if you see things the
     same as I do and, if so, what our relationship has that is missing in your re-
     lationship to your supervisor.
CLIENT: Well, for one thing, you listen to me, and I don't think she does. On
     the other hand, I listen pretty carefully to you, but I don't think I listen to
     her at all, and she probably knows it. I think she's dumb, and I guess I'm
     not hiding it from her.

The review of the relationship helps the client focus more specifically on his
relationship to his supervisor.
     Here is another example. Lee, a 38-year-old trainer in a counselor
training program, is talking to Carlos, 25, one of the trainees.

LEE: Carlos, I'm a bit bothered about some of the things that are going on be-
     tween you and me. When you talk to me, I get the feeling that you are being
     very careful. You talk slowly — you seem to be choosing your words, some-
     times to the point that what you are saying sounds almost prepared. You
     have never challenged me on anything in the group. When you talk most in-
     timately about yourself, you seem to avoid looking at me. I find myself giv-
     ing you less feedback than I give others. I've even found myself putting off
     talking to you about all this. Perhaps some of this is my own imagining, but
     I want to check it out with you.
CARLOS: I've been putting it off, too. I'm embarrassed about some of the
     things I think I have to say.

Carlos goes on to talk to Lee about his misgivings. He thinks that Lee is
domineering in the training group and plays favorites. He has not wanted to
bring it up because he feels that his position will be jeopardized. But now
that Lee has made the overture, he accepts the challenge.

Here-and-now immediacy.   Here-and-now immediacy refers to your ability to discuss with clients what is happening between you in the here and now of any given transaction. It is not the entire relationship that is being considered, but only this specific interaction or incident. In the following example, the helper, a 43-year-old woman, is a counselor in a church-related human service center. Agnes, a 49-year-old woman who was recently widowed, has been talking about her loneliness. Agnes seems to have withdrawn quite a bit, and the interaction has bogged down.

COUNSELOR: I'd like to stop a moment and take a look at what's happening right now between you and me.
AGNES: I'm not sure what you mean.
COUNSELOR: Well, our conversation today started out quite lively, and now it seems rather subdued to me. I've noticed that the muscles in my shoulders have become tense and that I feel a little flush. I sometimes tense up that way when I feel that I might have said something wrong. It could be just me but I sense that things are a bit strained between us right now.
AGNES (hesitatingly):   Well, a little.

Agnes goes on to say how she resented one of the helper's remarks early in the session. She thought that the counselor had intimated that she was lazy. Agnes knows that she isn't lazy. They discuss the incident, clear it up, and move on.

The purpose of here-and-now immediacy is to strengthen the working alliance. Research has shown that too much support can actually weaken the working alliance (*see* Kivlighan, 1990). Immediacy is a way of balancing support with challenge (Kivlighan & Schmitz, 1992; Tryon & Kane, 1993).

Self-involving statements.   Self-involving statements are present-tense, personal responses to the client (*see* Robitschek & McCarthy, 1991). They can be positive in tone; for example, "I like the way you've begun to show initiative both in discussion and outside. I thought that telling your boss a bit about your past showed guts." This self-involving remark is also a challenging statement, because the implication is "Keep it up." Clients tend to appreciate positive self-involving statements. In fact, "During the initial interview, the support and encouragement offered through the counselor's positive self-involving statements may be especially important because they put clients at ease and allay their anxiety about beginning counseling" (Watkins & Schneider, 1989, p. 345).

Negative self-involving statements are much more directly challenging in tone. Rogers, the dean of client-centered therapy, recounts the following incident.

> I am quite certain even before I stopped carrying individual counseling cases, I was doing more and more of what I would call confrontation. That is, confrontation of the other person with my feelings. . . . For example, I recall a client

with whom I began to realize I felt bored every time he came in. I had a hard time staying awake during the hour, and that was not like me at all. Because it was a persisting feeling, I realized I would have to share it with him. I had to confront him with my feeling and that really caused a conflict in his role as a client. . . . So with a good deal of difficulty and some embarrassment, I said to him, "I don't understand it myself, but when you start talking on and on about your problems in what seems to me a flat tone of voice, I find myself getting very bored." This was quite a jolt to him and he looked very unhappy. Then he began to talk about the way he talked and gradually he came to understand one of the reasons for the way he presented himself verbally. He said, "You know, I think the reason I talk in such an uninteresting way is because I don't think I have ever expected anyone to really hear me." . . . We got along much better after that because I could remind him that I heard the same flatness in his voice I used to hear. (Landreth, 1984, p. 323)

Rogers's self-involving statement, genuine but quite challenging, helped the client move forward. This story also points the direction in which Rogers was moving.

Situations Calling for Immediacy

Part of skilled helping is knowing when to use any given communication skill. The skill of immediacy can be most useful in the following situations:

- When a session is directionless and it seems that no progress is being made: "I feel that we're bogged down right now. Perhaps we could stop a moment and see what we're doing right and what's going wrong."
- When there is tension between helper and client: "We seem to be getting on each other's nerves. It might be helpful to stop a moment and clear the air."
- When trust seems to be an issue: "I see your hesitancy to talk, and I'm not sure whether it's related to me or not. It might still be hard for you to trust me."
- When there is "social distance" between helper and client in terms of social class or widely differing interpersonal styles: "There are some hints that the fact that I'm black and you're white is making both of us a bit hesitant."
- When dependency seems to be interfering with the helping process: "You don't seem willing to explore an issue until I give you permission to do so. And I seem to have let myself slip into the role of permission giver."
- When counterdependency seems to be blocking the helping relationship: "It seems that we're letting this session turn into a struggle between you and me. And, if I'm not mistaken, both of us seem to be bent on winning."

- When attraction is sidetracking either helper or client: "I think we've liked each other from the start. Now I'm wondering whether that might be getting in the way of the work we're doing here."

Immediacy — both in counseling and in everyday life — is a difficult, demanding skill. It is difficult, first of all, because the helper, without becoming self-preoccupied and without "psyching out" the client, needs to be aware of what is happening in the relationship and have enough psychological distance to catch difficult moments as they happen and react to them without trivializing the helping process. Second, immediacy often demands competence in all the communication skills discussed to this point — attending, listening, empathy, advanced empathy, and self-disclosure. Third, the helper must move beyond the MUM effect and challenge the client even though he or she is reluctant to do so. The helper needs guts. It is clear that people get into trouble in their day-to-day relationships because they do not have the skills — or perhaps the courage — needed to engage in either general-relationship or here-and-now immediacy. If this is the case, immediacy within sessions can help clients develop immediacy in real-life situations.

SUMMARIZING: PROVIDING FOCUS AND CHALLENGE

The communication skills of attending, listening, empathy, probing, and challenging need to be orchestrated in such a way as to help clients focus their attention on issues that make a difference. The ability to summarize and to help clients summarize the main points of a helping interchange or session is a skill that can be used to provide both focus and challenge. Brammer (1973) listed a number of goals that can be achieved by judicious use of summarizing: "warming up" the client, focusing scattered thoughts and feelings, bringing the discussion of a particular theme to a close, and prompting the client to explore a theme more thoroughly. Often, when scattered elements are brought together, the client sees the bigger picture more clearly. Thus, summarizing can lead to new perspectives or alternative frames of reference. In the following example, the client is a 52-year-old man who has revealed and explored a number of problems in living and is concerned about being depressed.

HELPER: Let's take a look at what we've seen so far. You're down — not just a normal slump — this time it's hanging on. You worry about your health, but you check out all right physically, so this seems to be more a symptom than a cause of your depression. There are some unresolved issues in your life. One that you seem to be stressing is the fact that your recent change in jobs has meant that you don't see much of your old friends anymore. Since you're single, you don't find this easy. Another issue — one you find painful and embarrassing — is your struggle to stay young. You don't like facing the fact that you're getting older. A third issue is the way you — to use your own word — "overinvest" yourself in work, so much so that when you finish a long-term project, suddenly your life is empty.

CLIENT (pauses):   It's painful to hear it all that baldly, but that about sums it up. I've suspected I've got some screwed-up values, but I haven't wanted to stop long enough to take a look at it. Maybe the time has come. I'm hurting enough.

HELPER: One way of doing this is by taking a look at what all this would look like if it looked better.

CLIENT: That sounds interesting, even hopeful. How would we do that?

The counselor's summary hits home — somewhat painfully — and the client draws his own conclusion. Care should be taken not to overwhelm clients with the contents of the summary. Nor should summaries be used to build a case against a client. Helping is not a judicial procedure. Perhaps the summary just given would have been more effective if the helper had also summarized some of the client's strengths. This would have provided a more positive context.

There are certain times when summaries prove particularly useful: at the beginning of a new session, when the session seems to be going nowhere, and when the client gets stuck.

At the beginning of a new session. When summaries are used at the beginning of a new session, especially when clients seem uncertain about how to begin, they prevent clients from merely repeating what has already been said before. They put clients under pressure to move on.

Liz began a session with a rather overly-talkative man with a summary of the main points of the previous session. This served several purposes. It showed the client that she had listened carefully to what he had said and that she had reflected on it after the session. Second, the summary gave the client a jumping-off point for the new session. It gave him an opportunity to add to or modify what was said. Finally, it placed the responsibility for moving on with the client. Clients might well need help to move on at times, but summaries give them the opportunity to exercise initiative.

Sessions that are going nowhere. A summary can be used to give focus to a session that seems to be going nowhere. One of the main reasons sessions go nowhere is that helpers allow clients to keep saying the same things over and over again instead of helping them either go more deeply into their stories or spell out the implications, especially the implication for action, of what they have said.

When a client gets stuck. Summaries can be used when clients seem to have exhausted everything they have to say about a particular issue and seem to be stuck. However, the helper does not always have to provide the summary. Often it is better to ask the client to pull together the major points. This helps the client own the process, pull together the salient points, and move on. Since this is not meant to be a way of testing clients, counselors should provide clients whatever help they need to stitch the summary together.

Summaries are often forms of challenge since they prevent a client from "just talking" and they apply pressure for more focus. They are invitations to get at more substantive issues and to spell out the implications of what has been said to this point.

PRINCIPLES UNDERLYING EFFECTIVE CHALLENGES

All challenges should be permeated by the spirit of the client-helper relationship values discussed in Chapter 3: they should be caring (not power games or put-downs), genuine (not tricks or games), designed to increase rather than decrease the self-responsibility of the client (not expressions of helper control), and pragmatic (not endless interpretations and the search for the ultimate insight). Clearly, challenging well is not a skill that comes automatically. It needs to be learned and practiced. The following principles constitute some basic guidelines.

Keep the goals of challenging in mind.   Challenge must be integrated into the entire helping process. Keep in mind that the goal is to help clients develop the kinds of alternative perspectives needed to clarify problem situations and to get on with the other steps of the helping process. Are the insights relevant rather than dramatic? To what degree do the new perspectives developed sit on the verge of action?

Encourage self-challenge.   Invite clients to challenge themselves, and give them ample opportunity to do so. You can provide clients with probes and structures that help them engage in self-challenge. In the following excerpt, the counselor is talking to a man who has discussed at length his son's ingratitude. There has been something cathartic about his complaints, but it is time to move on.

COUNSELOR: People often have blind spots in their relationships to others, especially in close relationships. I wonder whether you are beginning to see any blind spots in your relationship with your son.
CLIENT: Well, I don't know. . . . I guess I don't think about that very much. . . . Hmm. . . A thought just struck me. Let's say that, like me, he was sitting someplace with a counselor. What would he be saying about me? . . . I think I know what some of the things would be.

Alternatively, the counselor might have asked this client to list three things he thinks he does right and three things that need to be reconsidered in his relationship with his son. The point is to be inventive with the probes and structures you provide to help clients challenge themselves.

Earn the right to challenge.   Berenson and Mitchell (1974) maintained that some helpers don't have the right to challenge others, since they do not fulfill certain essential conditions. Here are some of the factors that earn you the right to challenge:

- *Develop a working relationship.* Challenge only if you have spent time and effort building a relationship with your client. If your rapport is poor or you have allowed your relationship with the client to stagnate, then deal with the relationship.
- *Make sure you understand the client.* Effective challenge flows from accurate understanding. Only when you see the world through the client's eyes can you begin to see what he or she is failing to see.
- *Be open to challenge yourself.* Don't challenge unless you are open to being challenged. If you are defensive in the counseling relationship or in your relationship with supervisors, do not expect your clients to set aside their defensiveness.
- *Work on your own life.* Berenson and Mitchell claimed that only people who are striving to live fully according to their value system have the right to challenge others, for only such persons are potential sources of human nourishment for others. In other words, helpers should be striving to develop physically, intellectually, socially, and emotionally.

In summary, ask yourself: What is there about me that will make clients willing to be challenged by me?

Be tentative but not apologetic in the way you challenge clients.   Tentative interpretations are generally viewed more positively than absolute interpretations (*see* Jones & Gelso, 1988). The same challenging message can be delivered in such a way as to invite the cooperation or arouse the resistance of the client. Deliver challenges tentatively, as hunches that are open to discussion rather than as accusations. Challenging is not an opportunity to browbeat clients or put them in their place. On the other hand, challenges that are delivered with too many qualifications — either verbally or through the helper's tone of voice — sound apologetic and can be easily dismissed by clients. I was once working in a career-development center. As I listened to one of the clients, it soon became evident that one reason he was getting nowhere was that he engaged in so much self-pity. When I shared this observation with him, I overqualified it. It came out something like this:

HELPER: Has it ever, at least in some small way, struck you that one possible reason for not getting ahead, at least as much as you would like, could be that at times you tend to engage in a little bit of self-pity?

I think it was almost that bad. I still remember his response. He paused, looked me in the eye and said: "A little bit of self-pity? . . . I *wallow* in self-pity." We moved on to explore what he might do to decrease his self-pity.

Build on successes.   High-level helpers do not urge clients to place too many demands on themselves all at once. Rather, they help clients place

reasonable demands on themselves and in the process help them appreciate
and celebrate their successes. In the following example, the client is a boy in
a detention center who is rather passive and allows the other boys to push
him around. Recently, however, he has made a few minor attempts to stick
up for his rights in the counseling group. The counselor is talking to him
alone after a group meeting.

COUNSELOR A: You're still not standing up for your own rights the way you
need to. You said a couple of things in there, but you still let them push you
around.

COUNSELOR B: Here's what I've noticed. In the group meetings you have begun to speak up. You say what you want to say, even though you don't say
it really forcefully. And I get the feeling that you feel good about that.
You've got power. You need to find ways of using it more effectively.

The first counselor does not reinforce the client for his accomplishment; the
second does.

Be specific.   Specific challenges hit the mark. Vague challenges get lost.
Clients don't know what to do about them. A statement such as, "You need
to pull yourself together and get on with it, don't you?" may satisfy some
helper need such as the ventilation of frustration but does little for clients.
On the other hand, a statement such as, "Let's see if we can list the resources you used when you took charge and completed that project" is
much more specific.

Challenge strengths rather than weaknesses.   Berenson and Mitchell
(1974) discovered that successful helpers tend to challenge clients' strengths
rather than their weaknesses. Individuals who focus on their failures find it
difficult to change their behavior. As Bandura (1986) has pointed out, clients
who regularly review their shortcomings tend to belittle their achievements,
to withhold rewards from themselves when they do achieve, and to live
with anxiety. All of this tends to undermine performance (*see* Bandura, 1986,
p. 339). Challenging strengths means pointing out to clients the assets and
resources they have but fail to use. In the following example, the helper is
talking to a woman in a rape crisis center who is very good at helping others
but who is always down on herself.

COUNSELOR: Ann, in the group sessions, you provide a great deal of support
for the other women. You have an amazing ability to spot a person in trouble and provide an encouraging word. And when one of the women wants
to give up, you are the first to challenge her, gently and forcibly at the same
time. When you get down on yourself, you don't accept what you provide
so freely to others. You are not nearly as kind to yourself as you are to
others.

The counselor helps her place a demand on herself to use her rather substantial resources on her own behalf.

Respect clients' values.   Challenge clients to clarify their values and to make reasonable choices based on them. Be wary of using challenging, even indirectly, to force clients to accept your values.

CLIENT (a 21-year-old woman who is a junior in college):   I have done practically no experimentation with sex. I'm not sure whether it's because I think it's wrong or whether I'm just plain scared.
COUNSELOR A: A certain amount of exploration is certainly normal. Perhaps some basic information on contraception would help allay your fears a bit.
COUNSELOR B: Perhaps you're saying that it's time to find out which it is.

Counselor A edges toward making some choices for the client, while Counselor B challenges her gently to find out what she really wants. Helpers can assist clients to explore the consequences of the values they hold, but this is not the same as attacking clients' values.

Deal with clients' defensiveness.   Do not be surprised when clients react strongly to being challenged. Help them share and work through their emotions. If they seem not to react externally to what you have said, elicit their reactions.

HELPER (after inviting a client to explore some self-destructive behavior):   I'm not sure how all this sounds to you.
CLIENT: I thought you were on my side. Now you sound like all the others. And I'm paying you to talk like this to me!

Even though the helper is tentative in his challenge, the client still reacts defensively. Try to get into a constructive dialogue with the disappointed or angered client. Remember, though, that an argument is not a dialogue. It may be that the client needs time to think about what you have said. Don't run from the interaction; on the other hand, don't insist on prolonging a conversation that is going nowhere.

In the long run, use your common sense. These are guidelines, not absolute prescriptions. As suggested in Chapter 1, take a no-formula approach. The more flexible and versatile you are, the more likely you are to be of benefit to your clients.

LINKING CHALLENGE TO ACTION

More and more theoreticians and practitioners are stressing the need to link insight to problem-managing action (*see* Westerman, 1989). Wachtel (1989) put it succinctly: "There is good reason to think that the really crucial insights are those closely linked to new actions in daily life and, moreover, that insights are as much a product of new experience as their cause" (p. 18). Ishiyama (1990), in a discussion of Morita therapy, talked about challenging and removing attitudinal blocks to constructive action. Certain attitudes —

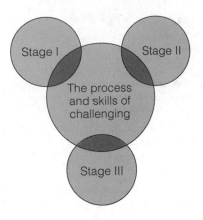

FIGURE 9-1
Challenge in the Helping Model

the need to reject and avoid inconvenient feelings, egocentric preoccupation, neglect of constructive desires, and so forth—sap energy that is needed for action. For instance, clients are encouraged to see anxiety and other unwanted feelings as part of human nature. Helping client "normalize" such feelings allows them to use their energy for more practical activities.

A few well-placed challenges might be all that clients need to move to constructive action. It is as if they were looking for someone to challenge them. Once challenged, they do whatever they need to manage their problems. Some clients are on the brink of challenging themselves and need only a nudge. Others, once they develop a few new perspectives, are off to the races. Still others have the resources—but not the will—to manage their lives better; they know what they need to do but are not doing it. A few nudges in the right direction help them overcome their inertia. I have had many one-session encounters that included a bit of listening, some empathy, and some new perspective that sent the client off on some useful course of action. On the other hand, if helping clients challenge themselves to develop new perspectives leads to nothing more than one profound insight after another but no action and behavioral change, then once more we are whistling in the wind.

Finally, as indicated in the first two chapters, social influence and challenging are not limited to Step I-B. Figure 9-1 indicates graphically that challenging or invitations to self-challenge can be part of any stage and any step.

How well am I doing the following as I try to help my clients?

General

- Becoming comfortable with the social-influence dimension of the helping role
- Incorporating challenge into my counseling style without becoming a confrontation specialist
- Using challenge wherever it is needed in the helping process
- Developing enough assertiveness to overcome the MUM effect

The Goals of Challenging

- Challenging clients to participate fully in the helping process
- Helping clients become aware of their blind spots in thinking and acting and helping them develop new perspectives
- Challenging clients to own their problems and unused potential
- Helping clients state problems in solvable terms
- Challenging clients' games, distortions, and excuses
- Inviting clients to explore the short- and long-term consequences of their behavior
- Helping clients move beyond discussion and inertia to action.

The Skills of Challenging

How effectively have I developed the communication skills that serve the process of challenging?

- *Information sharing* — giving clients needed information or helping them search for it to help them see problem situations in a new light and to provide a basis for action.
- *Advanced empathy* — sharing hunches with clients about their experiences, behaviors, and feelings to help them move beyond blind spots and develop needed new perspectives.
- *Helper self-disclosure* — sharing your own experience with clients as a way of modeling nondefensive self-disclosure and helping them move beyond blind spots.
- *Immediacy* — discussing aspects of your relationship with your clients in order to improve the working alliance.
- *Summarizing* — helping clients pull together pieces of their stories in order to see the bigger picture and move on in the helping process.

The Principles of Effective Challenging

When I do challenge, how effectively do I incorporate the following principles into my style?

- Keeping the goals of challenging in mind
- Inviting clients to challenge themselves
- Earning the right to challenge by
  - ☐ developing an effective working alliance with my client
  - ☐ working at seeing the client's point of view
  - ☐ being open to challenge myself
  - ☐ managing problems and developing opportunities in my own life
- Being tactful and tentative in challenging without being insipid or apologetic
- Being specific, developing challenges that hit the mark
- Challenging clients' strengths rather than their weaknesses
- Not asking clients to do too much too quickly
- Inviting clients to clarify and act on their own values, not mine

STEP I-C: LEVERAGE — HELPING CLIENTS WORK ON THE RIGHT THINGS

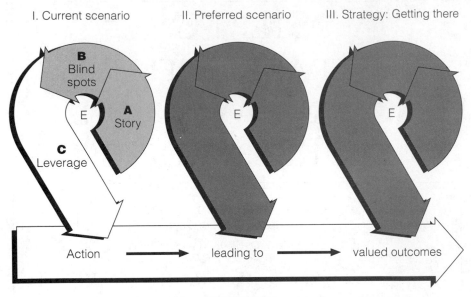

FIGURE 10-1
Step I-C: Leverage

Step I-C, called "leverage" and illustrated in Figure 10-1, involves helping clients choose which issues to work on during the helping process. This involves decision making on the part of the client. However, since decision making is involved in both the genesis of problems and the search for ways of managing them, this chapter starts with an overview of the decision-making process.

CLIENTS AS DECISION MAKERS

One of the reasons clients get into trouble in the first place is that they make poor decisions. Clients' stories are riddled with poorly made decisions. We need only to review our own experience to see how often our decisions or our failure to make decisions got us into trouble. In the helping process itself, there are many decision points. Clients must decide to come to a counseling interview in the first place; to talk about themselves; to return for a second session; to respond to the helper's empathy, probes, and challenges; to choose issues to work on; to determine what they want; to set goals; to develop strategies; to make plans; and to implement these plans. Deciding — versus letting the world decide for you — is at the heart of helping as it is at the heart of living.

Decision making in its broadest sense is the same as problem solving. Indeed, this book could be called a decision-making approach to helping. In this chapter, however, the focus is on decision making in a narrower sense — the internal (mental) action of identifying alternatives or options and choosing from among them. It is a commitment to do or to refrain from doing something; for example,

- "I have decided to discuss my career problems but not my sexual concerns."
- "I have decided to ask the courts to remove artificial life support from my comatose wife."
- "I have decided not to undergo chemotherapy."
- "I have decided to move into a retirement home."

The commitment can be to an internal action ("I have decided to get rid of my preoccupation with my ex-wife") or to an external action ("I have decided to confront my son about his drinking"). Decision making in the fullest sense includes the implementation of the decision; for instance, "I made a resolution to give up smoking, and I haven't smoked for three years." or "I decided that I was being too hard on myself so I took a week off work and just enjoyed myself."

Rational Decision Making

Traditionally, decision making has been presented as a rational, linear process involving information gathering, analysis, and choice. Here are the bare essentials of the decision-making process.

Information gathering.   The first rational task is to gather information related to the particular issue or concern. A patient who must decide whether to have a series of chemotherapy treatments needs some essential information. What are the treatments like? What will they accomplish? What are the side effects? What are the consequences of not having them? What would another doctor say? And so forth. And there is a whole range of ways in which she might gather this information: reading, talking to doctors, and talking to patients.

Analysis.   The next rational step is processing the information. This includes analyzing, thinking about, working with, discussing, meditating on, and immersing oneself in the information. Just as there are many different ways of gathering information, so there are many different ways of processing it. Effective information processing leads to a clarification and understanding of the range of possible choices. "Now, let's see, what are the advantages and disadvantages of each of these choices?" is a way of analyzing information. Effective analysis assumes that decision makers have criteria, whether objective or subjective, for comparing alternatives.

Making a choice.   Finally, decision makers need to make a choice; they need to commit themselves to some internal or external action that is based on the information and analysis. For example, "After thinking about it, I have decided to sue for custody of the children." The fullness of the choice includes an action: "I had my lawyer file the custody papers this morning." There are also rational rules that can be used to make a decision. For instance, one rule, stated as a question ("Will it get me what I want?") deals with the consequences of the decision.

Counselors help clients engage in rational decision making; they help clients gather information, analyze what they find, and then base action decisions on the analysis. This does happen, yet it is not the full story.

The Shadow Side of Decision Making: Choices in Everyday Life

In actuality, decision making in everyday life, and in counseling, is not the straightforward rational process just outlined (*see* Goslin, 1985; Heppner, 1989; Jeffery, 1989; Kaye, 1992; Schoemaker & Russo, 1990). Rather it is an ambiguous, highly complicated process; it has its shadow side. Headlee and Kalogjera (1988) saw some of the roots of this shadow side in childhood. Some children are allowed too much choice; others are given too few choices. Moreover, in the early years, distortions of choice evolve due to racial, ethnic, sexual, religious, and other prejudices. By the time the child becomes an adult, these are ingrained in the decision-making process and are not reflected on. In everyday life the decision making is often confused, covert, difficult to describe, unsystematic, and, at times, quite irrational. Many — if not most — of the decisions that are made everyday are based, not on reason, but on taste. A shadow-side analysis of decision making as it is actually practiced reveals a less-than-rational application of its three parts.

Information gathering.   Information gathering should lead to a clear definition of the matter to be decided. A client trying to decide whether to pursue a divorce needs information about that entire process. However, information gathering is practically never straightforward. For whatever reason, decision makers are often complacent and engage in an inadequate search; they get too much, too little, inaccurate, or misleading information; the search for information and the information itself are clouded with emotion. The client trying to decide whether to proceed with therapy may have already made up his or her mind and therefore not be open to confirming or disconfirming information. Since full, unambiguous information is never available, all decisions are at risk. In fact, there is no such thing as completely objective information. All information, especially in decision making, is received by the decision maker and takes on a subjective cast. In view of all this, Ackoff (1974) called human problem solving "mess management" (p. 21).

> Eloise wanted to make a decision whether to marry her companion or not. One of the obstacles was conflicting careers. She didn't know whether she'd be in the same career five years from now; neither did he. She knew little about his

past and thought that it didn't matter. She liked him now. Her companion knew that she was a non-practicing Catholic, but knew little about how her Catholicism affected her now or how it would affect them in the future, especially if they had children. Since religion was not currently an issue, they did not explore these issues.

There were many other things they did not know about themselves and each other. They eventually did marry, but the marriage lasted less than a year.

Granted, clients' stories are never complete, and information will always be partial and open to distortion. Yet, though counselors cannot help clients make information gathering perfect, they can help them make it adequate for problem management and opportunity development.

Processing the information.   Since it is impossible to separate the decision from the decision maker, the processing of information is as complex as the person making the decision. Factors affecting the analyzing of information include clients' feelings and emotions; their values-in-use, which often differ from their espoused values; their assumptions about how things work; and their level of motivation. There is no such thing as full, objective processing of gathered information. Poorly gathered information is often subjected to further mistreatment. Clients, because of their biases, focus on bits and pieces of the information they have gathered rather than the full picture. Furthermore, few clients have the time or patience to spell out all possible choices related to the issue at hand, together with the pros and cons of each. That is why most decisions are based, not on evidence, but on taste.

> Jamie was in a high-risk category for AIDS because of occasional drug use and sexual promiscuity. When he was arrested once for drug use, he had to attend a couple of sessions on AIDS awareness. He listened to all the information, but he processed it poorly. These were problems for "other people." He engaged in risky behavior "only occasionally." He was sure that his sexual partners were "clean." One or two "mistakes" were not going to do him in. He knew others who engaged in much riskier behavior than he and "nothing had happened to them." He'd be "more careful," though it was not clear what this meant in terms of behavior. He was in good health and, as he said, "healthy people can take a lot."

Jamie distorted information and rationalized away most of the risk of his current lifestyle.

Up to a point, counselors can help clients overcome inertia and biases and tackle the work of analysis. If a client says his values have changed, but he still automatically makes decisions based on his former values, then he can be challenged to get his new values into his decision making. One client, trying to make a decision about a career change, kept moving toward options in the helping professions even though he had become quite interested in business. There was something in him that kept saying, "You have to choose a helping profession. Otherwise you will be a traitor." The counselor helped him see his bias. He first became a consultant, then a manager, then a senior manager. But in the end, he had to assuage his conscience by noting both to himself and to others that "running a successful business is an important contribution to society."

Choice and execution.   A host of things can happen at the point of decision — the point of commitment — and at the implementation stage to make decision making an unpredictable process. Decision makers sometimes

- skip the analysis stage and move quickly to choice.
- ignore the analysis and base the decision on something else entirely. In such cases, the analysis was nothing but a sham because the decision criteria, however covert, were already in place.
- engage in what Janis and Mann (1977) called defensive avoidance. That is, they procrastinate, attempt to shift responsibility, or rationalize delaying a choice.
- confuse confidence in decision making with competence.
- panic and seize upon a hastily contrived solution that gives promise of immediate relief. The choice may work in the short term but have negative long-term consequences.
- are swayed by a course of action that is most salient at the time or by one that comes highly recommended, even though it is not right for them.
- let enthusiasm and other emotions govern their choices.
- announce a choice, whether to themselves or others, but then do nothing about it.
- translate the decision into action only half-heartedly.
- decide one thing but do another.

The fact that choices do not necessarily make life easier, as Goslin (1985) pointed out so well, but more difficult for self and for others explains a great deal of the shadow side of decision making. The "economics" of choosing are often poor. It is clear that counselors cannot help clients avoid all the pitfalls involved in making decisions, but they can help clients minimize them.

In summary, then, pure-form rational, linear decision making has probably never been the norm in human affairs. Decision making goes on at more than one level. There is, as it were, the rational decision-making process in the foreground and an emotional decision-making process in the background. Gelatt (1989) proposed an approach to decision making that factors in these realities: "What is appropriate now is a decision and counseling framework that helps clients deal with change and ambiguity, accept uncertainty and inconsistency, and utilize the nonrational and intuitive side of thinking and choosing" (p. 252). Positive uncertainty means, paradoxically, being positive (comfortable and confident) in the face of uncertainty (ambiguity and doubt): feeling uncertain about the future and positive

about the uncertainty. As we will see, stages II and III provide methodologies that clients can use to make decisions, explore their consequences, and act on them.

THE GOALS OF STEP I-C

Since Step I-C is both a stage-specific task and a process that permeates all
the stages of the helping process, its goals reflect both. Furthermore, even
though the following goals can be distinguished conceptually, in practice
they intermingle.

Helping clients become more effective decision makers.   This is obviously
a process goal: it relates to the entire helping process. Indeed, since self-
responsibility is a key helping value, then helping clients not only make
good decisions but become better decision makers is not an amenity but a
necessity.

Helping clients screen problems and opportunities.   Clients need to be
helped judge whether helping or therapy is what they need at this point.
Not everyone needs therapy and not every problem merits the kind of attention that counseling and therapy provide. The first-stage specific goal, then,
is to decide whether or not to continue. If the story being told has little
substance, then a decision may be made to terminate the relationship or
probe for more substantive issues.

Helping clients work on the right things through leverage.   Since all the
concerns in a complex problem/opportunity situation cannot be dealt with
at once, another goal is to help clients establish some priorities. Even problems that are told as single-issue stories ("I'd like to get rid of these headaches") may become multiple-issue problem situations once examined. For
instance, the headaches are a symptom of overwork, poor interpersonal
relationships, self-esteem problems, financial concerns, and an uncon-
trolled temper. Effective counselors help clients work on problems and opportunities that will make a difference. This is the search for leverage. The
helper asks, "What can I do to help this client get the most out of his or her
investment?" Once a client chooses a problem for attention, the helper uses
all the communication skills outlined earlier to help the client explore the
problem and develop new perspectives on it. The client must then be moved
to see what he or she would like to do about it in terms of a preferred scenario, action strategies, and implementation of these strategies.

Helping clients stay focused throughout the helping process.   A broader
goal is to help clients focus on the right things — not just the right issues, but
the right kind of relationship with the helper, the right goals, the right strategies for achieving those goals, and the right actions both within the helping
encounters and in day-to-day life. In this sense, step I-C is not a step but a

process that deals with the economics of helping. Helpers need to ask themselves, "Am I adding value through each of my interactions with this client?" Clients need to be helped to ask themselves, "Am I spending my time well? Do the decisions I am making have the potential of adding value to my life?"

We now take a closer look at two of these goals — screening and working on the right things.

SCREENING: THE INITIAL SEARCH FOR LEVERAGE

Helping is expensive in terms of financial and psychological costs. It should not be undertaken lightly. Relatively little is said in the literature about screening, deciding whether any given problem situation or opportunity deserves attention. The reasons are obvious. Helpers-to-be are rightly urged to take their clients and their clients' concerns seriously. They are also urged to adopt an optimistic attitude, an attitude of hope, about their clients. Finally, they are schooled to take their profession seriously and are convinced that their services can make a difference in the lives of clients. For these and other reasons, the first impulse of the average counselor is to try to help clients — no matter what the problem situation might be.

There is something very laudable about this. It is rewarding to see helpers prize people and express interest in their concerns. It is rewarding to see helpers put aside the almost instinctive tendency to evaluate and judge others and to offer their services to clients just because they are human beings. However, like other professions, helping can suffer from the "law of the instrument." Given a hammer, a child soon discovers that almost everything needs hammering. Once equipped with the models, methods, and skills of the helping process, helpers can see all human problems as needing their attention. In fact, in many cases counseling may be a useful intervention and yet be a luxury whose expense cannot be justified. The problem-severity formula discussed in Chapter 7 is a useful tool for screening.

Under the term *differential therapeutics*, Frances, Clarkin, and Perry (1984) discussed ways of fitting different kinds of treatment to different kinds of patients. They also discussed the conditions under which the best option is no treatment. In the no-treatment category they included clients who have a history of treatment failure or who seem to get worse from treatment, criminals trying to avoid or diminish punishment by claiming to be suffering from psychiatric conditions — as they say, "We may do a disservice to society, the legal system, the offenders, and ourselves if we are too willing to treat problems for which no effective treatment is available" (p. 227); patients with malingering or fictitious illness; people who are chronic nonresponders to treatment; clients likely to improve on their own; healthy clients with minor chronic problems; and reluctant and resistant clients who

refuse treatment. Although one might dispute some of the categories proposed, a decision needs to be taken in each case and the possibility of no
treatment does deserve serious attention.

The no-treatment option can do a number of useful things. It can inter-
rupt helping sessions that are going nowhere or that are actually destructive; it can keep clients and helpers from wasting time, effort, and money; it
can delay help until clients are ready or provide clients with opportunities to
discover that they can do without treatment; it can consolidate gains from
previous treatments; it can keep helpers and clients from playing games
with themselves and one another; and it can provide motivation for clients
to find help in their own daily lives. However, a no-treatment decision on
the part of helping professionals is countercultural and therefore difficult to
make.

It goes without saying that screening, or helping clients screen their
stories, can be done in a heavy-handed way. Statements such as "Your concerns are actually not that serious." or "You should be able to work that
through without help." or "I don't have time for problems as simple as that."
are simply not useful. Whether such sentiments are expressed or implied,
they obviously indicate a lack of respect and constitute a caricature of the
screening process. Helpers are not alone in grappling with this problem.
Physicians face clients day in and day out with problems that run from the
life threatening to the inconsequential. Statistics suggest that more than half
of the people who go to physicians have nothing physically wrong with
them. Consequently, physicians have to find ways to screen their patients'
complaints. I am sure that the best find ways to do so that preserve the
dignity of their patients.

Experienced helpers, because they are empathic, stay close to the experience of clients and develop hypotheses about both the substance of a
client's problems and the client's commitment and then test these hypotheses in a humane way. If clients' problems seem inconsequential, they
probe for more substantive issues. If clients seem reluctant, resistant, and
unwilling to work, they challenge clients' attitudes and help them work
through their resistance. In both cases they realize that there may come a
time, and it may come fairly quickly, to judge that further effort is uncalled
for because of lack of results. It is better, however, to help clients make such
decisions themselves or challenge them to do so. In the end, the helper
might have to call a halt, but his or her way of doing so should reflect basic
counseling values.

LEVERAGE: WORKING ON ISSUES THAT MAKE A DIFFERENCE

Clients often need help to get a handle on complex problem situations. A 41-
year-old depressed man with a failing marriage, a boring job, deteriorating
interpersonal relationships, health concerns, and a drinking problem

cannot work on everything at once. Priorities need to be set. Put bluntly, the question is: Where is the biggest payoff? Where should the limited resources of both client and helper be invested? If the problem situation has many dimensions, the helper and client are faced with the problem of where to start.

> Kiernan, a woman in her mid-thirties, is referred to a neighborhood mental health clinic by a social worker. During her first visit she pours out a story of woe both historical and current — brutal parents, teenage drug abuse, a poor marriage, unemployment, poverty, and the like. Kiernan, is so taken up with getting it all out that the helper can do little more than sit and listen.

Where is Kiernan to start? How can the time, effort, and money invested in helping provide a reasonable return to her? In the broadest sense, what are the economics of helping?

Some Principles of Leverage

The following principles of leverage — a reasonable return on the investment of the client's and helper's resources — serve as guidelines for choosing issues to be worked on. These principles overlap; more than one might apply at the same time.

- If there is a crisis, first help the client manage the crisis.
- Begin with the problem that seems to be causing pain for the client.
- Begin with issues the client sees as important.
- Begin with some manageable sub-problem of a larger problem situation.
- Begin with a problem that, if handled, will lead to some kind of general improvement in the client's condition.
- Focus on a problem for which the benefits will outweigh the costs.

Underlying all these principles is an attempt to make clients' initial experience of the helping process rewarding so that they will have the incentives they need to continue to work. Examples of the use and abuse of these principles follow.

If there is a crisis, first help the client manage the crisis.   Crisis intervention is sometimes seen as a special form of counseling (Baldwin, 1980; Janosik, 1984). It can also be seen as a rapid application of the three stages of the helping process to the most distressing aspects of a crisis situation.

> *Principle violated*: A resident-hall assistant (RA), a graduate student, meets an obviously agitated undergraduate student, Zachary, in the hall. Zachary says that he has to talk to him. They go to the RA's office to talk. As Zachary sits down, he is literally shaking. He finds it extremely difficult to talk. The prob-

lem is a case of attempted homosexual rape. Once Zachary blurts it out, the RA begins asking questions such as "Has this ever happened before?" "Do you think you did anything to make him think you were interested?" "How do you feel about your own sexual orientation?" "Do you think you should press charges?" and the like. Zachary talks with the RA for a few minutes and then explodes: "Why are you asking me all these silly questions?" He stalks out and goes to a friend's house.

Principle used: Seeing his agitation, his friend says, "Good grief, Zach, you look terrible. Come in. What's going on?" He listens to Zachary's account of what has just happened, interrupting very little, merely putting in a word here and there to let his friend know he is with him. He sits with Zachary when he falls silent or cries a bit, and then slowly and reassuringly "talks Zach down," engaging in an easy dialogue that gives his friend an opportunity gradually to find his composure again.

The friend's instincts are much better than those of the RA. He does what he can to defuse the immediate crisis.

Begin with the problem that seems to be causing pain.   Clients often come for help because they are hurting even though they are not in crisis. Their hurt, then, becomes a point of leverage. Their pain also makes them vulnerable. This means they are open to influence from helpers. If it is evident that they are open to influence because of their pain, seize the opportunity, but move cautiously. Their pain may also make them demanding. They can't understand why you cannot help them get rid of the pain immediately. This kind of impatience may put you off, but it needs to be understood. Such clients are like patients in the emergency room, each one seeing himself or herself as needing immediate attention. Their demands for immediate relief may well signal a self-centeredness that is part of their character and therefore part of the broader problem situation. It may be that, in your eyes, their pain is self-inflicted; ultimately you may have to challenge them on this. But pain, whether self-inflicted or not, is still pain. Part of your respect for clients is your respect for them as vulnerable.

Principle violated: Rob, a man in his mid-twenties, comes to a counselor in great distress because his wife has just left him. The counselor sees quickly that Rob is an impulsive, self-centered person with whom it would be difficult to live. The counselor immediately challenges him to take a look at his interpersonal style and the ways in which he alienates others. Rob seems to listen, but he does not return.

Principle used: Rob goes to a second counselor who also sees a number of clues indicating self-centeredness and a lack of maturity and discipline. However, she listens carefully to his story, even though it is one-sided. Instead of adding to his pain by making him come to grips with his selfishness, she focuses on the future. Of course, his wife's return is the most important part of a better future. Since this is the case, the counselor also helps him describe in some detail what life would be like once they got back together. His hurting provides the incentive for working with the counselor on how he needs to change to get what he wants.

The helper's focusing on a better future responds to Rob's need. This future, then, becomes the stepping stone to challenge. What is Rob willing to do to create the future he says he wants?

Begin with issues the client sees as important and is willing to work on. The frame of reference of the client is a point of leverage. Given the client's story, you may think that he or she has not chosen the most important issues for initial consideration. However, helping clients work on issues that are important to them sends an important message: "Your interests are important to me."

> *Principle violated*: A woman comes to a counselor complaining about her relationship with her boss. She believes that he is sexist, and there are hints of sexual overtures. After listening to her story, the counselor believes that she probably has some leftover developmental issues with her father that affect her attitude toward older men. He pursues this line of thinking with her. The client is confused and feels put down. When she does not return for a second interview, the counselor says to himself that she is not motivated to do something about her problem.

> *Principle used*: The same women then talks to a pastoral counselor about her concerns with her boss. The counselor listens and picks up indications that the woman has strong feelings and unresolved issues about her father that she has not recognized. However, for the present the counselor does not address these issues. Instead, she helps the woman deal with her immediate concerns. The counselor will look for an opportunity to help the woman come to grips with those developmental issues later on.

It may well be that a client's immediate frame of reference needs broadening. It may be clear to the helper that the client is skirting around the real issues. Effective helpers start with the frame of reference of the client and then, using the challenging skills discussed in Chapters 8 and 9, help the client broaden this frame to include issues that make a difference as quickly as possible.

Begin with some manageable sub-problem of a larger problem situation. Large, complicated problem situations remain vague and unmanageable. Dividing a problem into manageable bits can provide leverage. Most larger problems can be broken down into smaller, more manageable sub-problems.

> *Principle violated*: Lisa, a single woman in her mid-thirties, has not held a job for more than a year or two. She is presently seeking career counseling, a bit reluctantly. Though intelligent and talented, she has been laid off several times. Two of the companies, when laying her off, talked about "general cut-

backs," but a third cited "a lack of cooperation with other workers" and had the documentation needed for the layoff. Lisa is seen by most people as "difficult." Her abrasive ways tend to alienate people who come in close contact with her. In two sessions the counselor, too, finds her an abrasive person who has little insight into how demanding she is. The counselor listens to her concerns and then says something like this: "Well, this is not really a career problem. This is an entire *lifestyle* problem. Your whole approach to people is going to stand in the way of your getting the kind of job you want, especially in these days of downsizing. I'm afraid that we have a lot of work to do." Lisa does not return to see this counselor.

Principle used: A couple of weeks pass and Lisa screws up her courage to see another counselor. The second counselor, too, experiences her as quite abrasive, but it is clear to her that Lisa is not in touch with her interpersonal style. This counselor does not immediately deal with Lisa's abrasiveness. Rather, she suggests that one practical step is to help her see how she comes across in job interviews. The counselor videotapes a couple of role-playing sessions and helps Lisa critique what she sees. Even though Lisa does not like what she sees, she finds this quite useful and begins to establish a working alliance with the counselor. The videotapes serve as a bridge to more sensitive issues.

Lisa's entire interpersonal style is not an immediately manageable problem. The first counselor asks for too much too soon and alienates Lisa with his confrontation. In the second case the counselor starts with a felt need of the client—getting a job—and focuses on one dimension of the job-seeking process—the interview. In true form, Lisa expresses some of her abrasiveness in the interview. Having Lisa critique her own performance in an interview is more manageable than dealing with the entire area of interpersonal style.

Move as quickly as possible to a problem that, if handled, will lead to some kind of general improvement. Some problems, when addressed, yield results beyond what might be expected. This is the spread effect.

Principle violated: Jeff, a single carpenter in his late twenties, comes to a community mental health center with a variety of complaints, including insomnia, light drug use, feelings of alienation from his family, a variety of psychosomatic complaints, and temptations toward exhibitionism. He also has an intense fear of dogs, something that occasionally affects his work. The counselor sees this problem as one that can be managed through the application of behavior modification methodologies. He and Jeff spend a fair amount of time in the desensitization of the phobia. Jeff's fear of dogs abates quite a bit and many of his symptoms diminish. But gradually the old symptoms re-emerge. His phobia, while significant, is not related closely enough to his major concerns to involve any kind of major spread effect.

Principle used: Jeff's major problems re-emerge. One day he becomes disoriented and bangs a number of cars near his job site with his hammer. He is overheard saying, "I'll get even with you." He is admitted briefly into a general hospital with a psychiatric ward. The immediate crisis is managed quickly and effectively. During his brief stay he talks with a psychiatric social worker. He feels good about the interaction and they agree to have a few sessions after he is discharged. In their talks, the focus turns to Jeff's isolation. In adolescence he learned that sex was bad. Since he had sexual feelings for both men and women, he solved his problem by staying away from close relationships with men and women. Jeff and the social worker go on to discuss the need for intimacy in his life and ways of getting back into the community. The intimacy of the counseling discussions themselves help a great deal and Jeff's symptoms begin to disappear.

Guilt about intimacy and the lack of intimacy in his life are high-leverage issues in Jeff's life. Dealing with intimacy and the associated guilt takes care of many of the other complaints in his life. Effective counselors help clients find these leverage themes and capitalize on the resources they have to deal with them.

Focus on a problem for which the benefits will outweigh the costs.   This is not an excuse for not tackling difficult problems. If you demand a great deal of work from yourself and from the client, then basic laws of behavior suggest that there be some kind of reasonable payoff for both of you.

> *Principle violated*: Margaret was married to Hector, a man who had been diagnosed to be HIV-positive. The fact that she and her recently-born son had tested negative helped cushion her shock. But she was having difficulty refocusing her relationship with Hector. He said that he had picked up the virus from a "dirty needle." She didn't even know that he had ever used drugs. The helper's approach focused on "personality reconstruction." His idea was that she needed to reconstruct parts of her personality to deal with this new situation. He told her that some of this would be painful because it meant looking at areas of her life that she had never reviewed or discussed. She wanted some practical help in re-orienting herself to her husband and to family life. After three sessions she did not return.

> *Principle used*: Margaret still searches for help. The doctor who is treating Hector suggests a self-help group for spouses, children, and companions of HIV-positive patients. There is a small weekly fee for the group. In the sessions Margaret learns a great deal about how to relate to someone who is HIV-positive. The meetings are very practical. The fact that Hector is not in the group helps. She begins to understand herself and her needs better. In the security of the group she explores mistakes she had made in relating to Hector. While she does not "reconstruct" her personality, she does learn how to live more creatively with her husband. She realizes that there will probably be a lot of anguish and pain in the future, but she also sees that she will be much better prepared to face that future.

Personality "reconstruction," even if possible, is a very costly proposition. It is not at all clear that Margaret needs such reconstruction. It is clear that the cost-benefit ratio is completely out of balance. Once more it is a question of a helper more committed to his theories than to the needs of his clients. Margaret gets the help she needs in the group and in sessions she and Hector have with a caring doctor.

A No-Formula Approach

These are principles, not formulas. Keeping the outcome in mind — helping the client get focused and the effective and efficient use of time and other resources at the service of problem management and opportunity development — is important. The principles of leverage, used judiciously and in combination, can help counselors shape the helping process to suit the needs and resources of individual clients. However, helpers must avoid using these very same principles to water down or retard the helping process. Ineffective helpers

- begin with the clients' framework but never get beyond it;
- fail to challenge clients to consider significant issues they are avoiding;
- allow clients' pain and discomfort to mask the roots of their problems;
- fail to help clients build on their successes;
- continue to deal with small, manageable problems and fail to help clients face more demanding problems in living;
- fail to help clients generalize learning in one area of life (for instance, self-discipline in a fitness program) to other, more difficult areas of living (for instance, self-control in closer interpersonal relationships).

In a word, ineffective helpers are afraid of making reasonable demands of clients or of influencing clients to make reasonable demands of themselves.

FOCUS AND LEVERAGE: THE LAZARUS TECHNIQUE

In a film on his multimodal approach to therapy, Lazarus (Lazarus, 1976, 1981; Rogers, Shostrom, & Lazarus, 1977) used a focusing technique I find useful. He asked the client to use just one word to describe her problem. She searched around a bit and then offered a word. For instance, a client might say, "Cloudy." Then he asked the client to put the word in a simple sentence that would describe her problem. The client might say, "My mind is cloudy and I can't think straight." He then asked for a more extended description of the issue. This client said something like this:

> When I say that my mind is cloudy and I can't think straight I mean that this is the reaction to what my boss told me last week. I have been working very hard. And learning a lot. Over the past two years I've received two awards for special projects. And all along I've gotten good ratings. But then the recession hit and they were getting rid of all sorts of people. I was let go because the unit I was in

was eliminated. But if I was supposedly so good, why didn't they find a way of keeping me? . . . I was shell-shocked. . . . I've been very confused . . . and in a fog. . . . I was loyal. . . . It's hit me very hard.

I'm not suggesting that you begin all or any session that way, but it is a simple way to bring a session into focus. This methodology can be used at any stage or step of the helping process. For instance, clients can be asked to use one word to describe what they want.

HELPER: Now that you have explored some of your concerns and against this background, tell me in just one word or short phrase what you want.
CLIENT: Let's see. . . . My family.
HELPER: All right, your family. Now put that in a sentence. Describe what you want in a short sentence.
CLIENT: I want my family and my family life back.

This client had achieved financial success in a business venture rather quickly. Success went to his head. He left his wife and three children and began living with a flashier woman. Now, some six years later, the shine was off the apple. He was successful but unhappy. He wanted his family back; he wanted family life as he once knew it again. This for him was a leverage point. How realistic this was and how it could be achieved was another issue. But the Lazarus technique provided focus.

STEP I-C AND ACTION

Like every other step of the helping model, Step I-C can act as a stimulus to client action. Helping clients find leverage means, concomitantly, helping them discover incentives for acting inside the helping sessions and outside. Helping clients identify and deal with high-leverage issues helps them move toward the little actions that precede the formal plan for constructive change. One client, a man in his early thirties, discussed all sorts of different problems and unused opportunities with a counselor. He skipped from one topic to another, usually settling on the issue that had caused him the most trouble the previous week. The counselor always listened attentively. At the beginning of the third session she said, "I have a hypothesis I'd like to explore with you." She went on to name what she saw as a thread that wove its way through most of the issues he discussed—his reluctance to commit himself. He was thunderstruck, because he instantly recognized the truth. In the next few weeks he began to explore his problems and concerns from this perspective. Time after time he saw himself withdrawing when he should have been committing himself, whether to a project or to a person. Gradually he began to commit himself in little ways. For instance, when an intimate woman friend of his began talking about where their relationship was going, he said, "There is something in me that wants to change the subject or run away. But no. Let's begin talking about the future and what future we might have together." At least he committed himself to opening a discussion about the relationship if not to the relationship itself. Discussing issues that make a difference can galvanize some clients into using resources that have lain dormant for years.

A Final Note on Stage I

Stage I is both a stage and a process. As a stage, it deals with clients' telling their stories, choosing issues to focus on, and clarifying these issues in terms of specific experiences, behaviors, and emotions. Effective helpers both support and challenge clients in this pursuit. As process, story telling, challenging, and the search for leverage belong in all the stages of the helping encounter. There is no time when further relevant elements of a story or a new part of the overall story cannot emerge. Both self-challenge and the challenges meted out by helpers need to permeate the entire process. Challenge makes the helping interviews different from the encounters in day-to-day life—and the helping sessions have to be different if they are to make a difference. The search for leverage must permeate every stage and step of the helping process. It makes no sense to spend time on things that don't make a difference.

EVALUATION
QUESTIONS FOR STEP I-C

How well am I doing the following?

- Observing clients' decision-making style and helping them not only to make good decisions but also to become better decision makers
- Helping clients focus on issues that have payoff potential for them
- Helping clients make critical decisions
- Maintaining a sense of movement and direction in the helping process
- Avoiding unnecessarily extending the problem identification and exploration stage
- Moving to other stages of the helping process as clients' needs dictate
- Encouraging clients to act on what they are learning
- Helping clients make their actions prudent and directional

PART FOUR

HELPING CLIENTS DEVELOP PROGRAMS FOR CONSTRUCTIVE CHANGE

The first chapter in Part 4, Chapter 11, is an introduction to some of the perspectives and skills that counselors need in order to help clients create better futures. With this as a base, we then move to Stages II and III, the most important parts of the helping model. It is here that counselors help clients develop and implement programs for constructive change. The payoff for identifying and clarifying problem situations and unused opportunities lies in doing something about them. The skills that helpers and clients need to do precisely this — engage in constructive change — are reviewed and illustrated in Stage II (Chapters 12 and 13) and Stage III (Chapters 14 through 17).

CHAPTER 11

PERSPECTIVES AND SKILLS FOR CONSTRUCTING A BETTER FUTURE

DEVELOPING PROGRAMS FOR
CONSTRUCTIVE CHANGE

The goal of the counselor is to help clients engage in constructive change. Stages II and III deal with programs for constructive change, which, as Carkhuff (1985) has noted, can have powerful effect:

> Program development lets us minimize and ultimately rise above our human limitations. It enables us to achieve goals we once only imagined reaching. And just as program development has made it possible for human beings to walk in space, it allows us to walk proudly on earth: strong, healthy, skilled, and productive. (from the Preface)

Program development together with the skills and attitudes needed to develop them are too often overlooked in the helping industry. Both helper and client need certain perspectives and skills to develop such programs — brainstorming possible goals, choosing specific goals, brainstorming implementation strategies, choosing implementation strategies, and then putting these strategies into a sequence with timelines for each activity — all at the service of managing the problems and developing the unused opportunities discovered and analyzed in Stage I. Chapter 11 reviews these perspectives and skills.

In Stages II and III, counselors help clients construct a better future for themselves and discover ways of making it a reality. For many clients the power of the helping process is found mainly in these two stages. The problem is that helpers are too often biased toward Stage I. That is, they spend an inordinate amount of time helping clients identify, explore, and clarify problem situations in the hope that this will lead to constructive change. They also spend an inordinate amount of time using a variety of theories to help clients develop insights into themselves and their problems in living and then stop short of helping clients translate these insights into programs for change. The real challenge of helping is not to help clients identify and clarify problem situations, however essential this might be, but to help clients manage them. Too many helper training programs begin and end with basic communications skills and the steps of Stage I. We shortchange helpers-to-be if we fail to offer them the tools they need to help clients develop programs for constructive change.

WHAT DO YOU WANT?

When President-elect Clinton made his first visit to Washington after the 1992 presidential election, he spoke with President Bush at the White House and then took a walk through an economically depressed area nearby. One woman began telling him the woes of the city. In response, Clinton asked, "What do you want?" The woman looked a little startled and then, recovering, said something like this: "We want jobs. We want safe streets. We want the drug pushers out of our neighborhood." Clinton asked a question that is not often asked by either politicians or helpers. It is a powerful, radically client-centered question. In Stages II and III, counselors must help clients ask themselves and answer the following two common-sense, but critical, questions: *"What do you want?"* and *"What do you have to do to get what you want?"* These two questions express the guts of Stages II and III in the language of the client. In more program-development terms, Stages II and III involve helping clients set problem-managing and opportunity-developing goals and find strategies for accomplishing them. This is the way forward. Goal setting and the development of strategies for implementing goals are the two key features of a program for constructive change. For many clients the notion of setting goals has no punch. On the other hand, being asked and then helped to answer the question, "What do you really want?" has a great deal of power.

GOAL SETTING: POWER AND PROBLEM

Goal setting — helping clients come to grips with what they want — is one of the principal ways in which clients can exercise their power, but it also constitutes a problem in the helping process. A look at both — first power, then problem.

Power: Benefits of Goal Setting

At their best, goals mobilize clients' resources at the service of action — they get people moving — and provide channels for wise action — they get clients headed in the right direction.

Goals as stimuli for action.   According to Locke and Latham (1984), helping clients set goals empowers them in four ways:

   1.   *Goals focus clients' attention and action.*   A counselor at a refugee center in London described Simon, a victim of torture in a middle-eastern country, to her supervisor as aimless and minimally cooperative in exploring the meaning of his brutal experience. Her supervisor suggested that she help Simon explore a better future. She started one session by asking, "Simon, if you could have one thing you don't have, what would it be?" Simon came back immediately, "A friend." During the rest of the session he was totally focused. What was uppermost in his mind was not the torture but the fact that he was so lonely in a foreign country. When he did talk about the torture it was in terms of his fear that torture had "disfigured" him, if not physically, then psychologically, thus making him unattractive to others.
   2.   *Goals mobilize clients' energy and effort.*   Clients who seem lethargic during the problem exploration phase often come to life when asked to

discuss possibilities for a better future. A patient in a long-term rehabilitation program who had been listless and uncooperative said to her counselor
after a visit from her minister: "I've decided that God and God's creation
and not pain will be the center of my life. This is what I want." This was the
beginning of a new commitment to the arduous program. She collaborated
more fully in exercises that helped her manage her pain. Clients with goals
are less likely to engage in aimless behavior. Goal setting is not just a mental
exercise. Many clients begin engaging in constructive change after setting
even broad or rudimentary goals.

   3. *Goals stated in specific terms increase persistence.*   Not only are clients
with goals energized to do something, but they tend to work harder and
longer. An AIDS patient who said that he wanted to be reintegrated into his
extended family managed, against all odds, to recover from five separate
hospitalizations in order to achieve what he wanted. He did everything
he could to buy the time he needed. Clients with clear and realistic
goals don't give up as easily as clients with vague goals or with no goals
at all.

   4. *Setting goals motivates clients to search for strategies to accomplish
them.*   Setting goals, a Stage-II task, leads naturally into a search for means
to accomplish them, a Stage-III task. Lonnie, a woman in her 70s, had been
described by her friends as "going down hill fast." After a heart-problem
scare that proved to be a false alarm, she decided that she wanted to live
until she died. She searched out ingenious ways of redeveloping a social
life, including remodeling her house and taking in two young women from
a local college as boarders.

   It is puzzling, then, to see counselors helping clients explore problem
situations and unused opportunities and then stopping short of asking
them what they want to replace what they have. As Bandura (1990) put it,
"Despite this unprecedented level of empirical support [for the advantages
of goal setting], goal theory has not been accorded the prominence it deserves in mainstream psychology" (p. xii). One of the main reasons that
counselors do not help clients develop realistic goals is that we do not train
them to do so.

Goals as channels of wise action.   The action orientation of the helping
model has been emphasized over and over again. But action needs direction. Goals provide direction and help make action prudent.

   A sense of direction. Setting goals, however informally, provides clients
with a sense of direction. People with a sense of direction:

- have a sense of purpose.
- live lives that are going somewhere.
- have self-enhancing patterns of behavior in place.
- focus on results, outcomes, and accomplishments.

- don't mistake aimless action for accomplishments.
- set goals and objectives, though they might not use these words.
- have a defined rather than an aimless lifestyle.

One study (Payne, Robbins, & Dougherty, 1991) showed that high-goal-directed retirees were more outgoing, involved, resourceful, and persistent in their social settings; and the low-goal-directed were more self-critical, dissatisfied, sulky, and self-centered. People with a sense of direction don't waste time in wishful thinking. Instead, wishes are translated into specific outcomes toward which they can work. Picture a continuum. At one end is the aimless person; at the other, a person with a sense of direction. Your clients might come from any point on the continuum. They may be at different points with respect to different issues — mature in seizing opportunities for education, for instance, but aimless in developing sexual maturity. All of us are marginal at one time or another. Helping clients establish and commit themselves to problem-managing goals is the principal task of Stage II.

Wisdom and prudence. The helping model described in this book encourages a bias toward action on the part of clients; however, this action needs to be directional and wise. Discussing and setting goals contributes to direction and wisdom. The following case began poorly but ends well.

> Harry was a sophomore in college who was admitted to a state mental hospital because of some bizarre behavior at the university. He was one of the disc jockeys for the university radio station. He came to the notice of college officials one day when he put on an attention-getting performance that included rather lengthy dramatizations of grandiose religious themes. In the hospital it was soon discovered that this quite pleasant, likable young man was actually a loner. Everyone who knew him at the university thought that he had a lot of friends, but in fact he did not. The campus was large, and his lack of friends had gone unnoticed.
>
> Harry was soon released from the hospital but returned weekly for therapy. At one point he talked about his relationships with women. Once it became clear to him that his meetings with women were perfunctory and almost always took place in groups — yet he had actually thought he had a rather full social life with women — Harry launched a full program of getting involved with the opposite sex. His efforts ended in disaster, because Harry had some basic sexual and communication problems. He also had serious doubts about his own worth and therefore found it difficult to make a gift of himself to others. He ended up in the hospital again.
>
> The counselor helped Harry get over his sense of failure by emphasizing what Harry could learn from the "disaster." With the therapist's help, Harry returned to the problem-clarification and new-perspectives part of the helping process and then established more realistic short-term goals regarding getting back into the community. The direction was the same — establishing a realistic social life — but the goals were now more prudent because they were bite sized. Harry attended socials at a local church where a church volunteer provided support and guidance.

Harry's leaping from problem clarification to action without taking time to discuss possibilities and set a direction in terms of reasonable goals was part of the problem rather than part of the solution. His lack of success with women actually helped him see his problem with women more clearly. There are two kinds of prudence—playing it safe is one, doing the wise thing is the other. Helping is venturesome. It is about wise choices rather than playing it safe.

Problem: Perception of Goal Setting as an Overly Rational Process

Despite the advantages of goal setting just outlined, the pursuit and accomplishment of goals sounds very rational, perhaps too rational, for some. Perhaps because Stage II is not part of the natural problem-managing process outlined in Chapter 2, not only do clients not instinctively set problem-managing and opportunity-developing goals, but counselors do not instinctively help them do so. In this book we talk about analyzing problem situations, establishing goals to manage them, and achieving these goals through the implementation of a plan. Because the real-life process of problem management and opportunity development is messier, helpers and clients object to this overly rational approach. There is a dilemma. Many clients need or would benefit from a rigorous application of the problem-management process. Yet they find it alien and resist its rationality and discipline. Some kind of compromise is needed. Perhaps Langer (1989) put it well when he suggested that clients should be goal-guided rather than goal-governed.

Helping clients explore what they want.   For some, the very term *goal setting* gets in the way. Therefore, helpers should use whatever language makes sense to clients to describe the process of Stages II and III. For instance, goal setting has a technological feel to it, but "helping clients explore what they want" does not. In the following example, the helper is talking to an unmarried woman who, for a number of years, has been working a full-time job and taking care of her mother who has Alzheimer's disease. Lately the woman has complained of vague physical problems and depression.

COUNSELOR: I think I know what you want for your mother, but I'm not sure what you want for yourself.
CLIENT: I guess I don't usually think about myself very much. But these aches and pains have made me much more self-centered than I usually am. Well, in some ideal world, I'd like a lot of things. I'd like some relief in taking care of my mother. It gets harder all the time. I'd like some time for a social life, but between work and mother there just isn't any. To tell you the truth, I'd like a vacation. I haven't had a real one for years. . . . On a personal level, I'd like some friendship. Even a male friend.

These are goals, but the counselor does not use the terms *goals* or *goal setting*. Later, he helps her sift through these and other goals, choose a package that has some realism to it, and find ways of accomplishing them. Of course, in some cases it is a question of helping clients determine not just what they want but also what they need. For instance, some clients need greater self-discipline even though they don't state this as one of their wants. Then goal setting becomes an even more challenging process.

Helping clients identify and clarify emerging goals.   Goals need not be set. Rather they can naturally emerge through the client-helper dialogue. The skills and methods of Stage I, used effectively, do not lead exclusively to the clarification of problems and opportunities. Possible goals and action strategies often come tumbling out too. Having been helped to clarify a problem situation through a combination of probing, empathy, and challenge, clients begin to see more clearly what they want and what they have to do to manage the problem. Some clients must first act in some way before they find out just what they want to do. Once goals begin to emerge, counselors can help clients clarify them and find ways of implementing them. A caution: if clients wait around until "something comes up" or try a lot of different solutions in the hope that one of them will work, then emergence can be a self-defeating process.

Explicit goal setting is not to be underrated. Some clients are looking for this kind of direction right from the start. If this is the case, then helping clients set goals, even quite specific goals, is a way of meeting their needs. Taussig (1987) showed that clients respond positively to goal setting even when problems are discussed and goals set very early in the counseling process. The best approach seems to be a client-centered, no-formula approach. All clients need focus and direction in managing problems and developing opportunities; what that focus and direction will look like will differ from client to client.

THE SHADOW SIDE OF GOAL SETTING

As suggested earlier, helpers sometimes become co-conspirators with clients in helping them avoid goal setting under whatever name. There are a number of reasons for this, though they are not often discussed in the helping sessions.

1.   Goal setting means that clients have to move out of the relatively safe harbor of discussing problem situations and of exploring the possible roots of these problems in the past and move into the uncharted waters of the future. This may be uncomfortable for client and helper alike.

2.   Clients who set goals and commit themselves to them move beyond the victim game. Victimhood and self-responsibility make poor bedfellows. For some clients and helpers, victimhood is the easier choice.

3.   Goal setting involves clients' placing demands on themselves, making decisions, committing themselves, and moving to action. It requires action. If I say, "This is what I want," then, at least logically, I must also say, "Here is what I am going to do to get it. I know the price and I am willing to pay it." Since this demands work and pain, clients will not always be grateful for this kind of help.

4.   Although liberating in many respects, goals also hem clients in. If a woman chooses a career, then it might not be possible for her to have the kind of marriage she would like. If a man commits himself to one woman, then he can no longer play the field.

Effective helpers know what lurks in the shadows of goal setting for themselves and for their clients. They are prepared to manage their own part of it and to help clients manage theirs.

PERSPECTIVES AND SKILLS FOR CREATING A BETTER FUTURE

At its best, counseling helps clients move into discovery mode. Discovery mode involves creativity and divergent thinking.

Creativity and Helping

One of the myths of creativity is that some people are creative and others are not. Clients, like the rest of us, can be more creative than they are. It is a question of finding ways to help them be so. Stages II and III help clients tap into their dormant creativity. A review of the requirements for creativity (*see* Cole & Sarnoff, 1980; Robertshaw, Mecca, & Rerick, 1978, pp. 118–120) shows, by implication, that people in trouble often fail to use whatever creative resources they might have. The creative person is characterized by:

- *optimism and confidence*, whereas clients are often depressed and feel powerless.
- *acceptance of ambiguity and uncertainty*, whereas clients may feel tortured by ambiguity and uncertainty and want to escape from them as quickly as possible.
- *a wide range of interests*, whereas clients may be people with a narrow range of interests or whose normal interests have been severely narrowed by anxiety and pain.
- *flexibility*, whereas clients may have become rigid in their approach to themselves, others, and the social settings of life.
- *tolerance of complexity*, whereas clients are often confused and looking for simplicity and simple solutions.

- *verbal fluency,* whereas clients are often unable to articulate their problems, much less their goals and ways of accomplishing them.
- *curiosity,* whereas clients may not have developed a searching approach to life or may have been hurt by being too venturesome.
- *drive and persistence,* whereas clients may be all too ready to give up.
- *independence,* whereas clients may be quite dependent or counter-dependent.
- *nonconformity or reasonable risk taking,* whereas clients may have a history of being very conservative and conformist, or may be people who get into trouble with others and with society precisely because of their particular brand of nonconformity.

A review of some of the principal obstacles to creativity reveals further problems. Innovation is hindered by:

- *fear,* and clients are often quite fearful and anxious.
- *fixed habits,* and clients may have self-defeating habits or patterns of behavior that may be deeply ingrained.
- *dependence on authority,* and clients may come to helpers looking for the "right answers" or be quite counterdependent (the other side of the dependence coin) and fight efforts to be helped with "Yes, but" and other games.
- *perfectionism,* and clients may come to helpers precisely because they are hounded by this problem and can accept only ideal or perfect solutions.

It is easy to say that imagination and creativity are most useful in Stages II and III, but it is another thing to help clients stimulate their own, perhaps dormant, creative potential. Therefore,

- assume that your clients have unused creative resources that can be brought to bear on problem management and opportunity development. Help them see that everything they do can be done in a slightly more inventive way.
- create the right atmosphere for creativity. Help clients relax and realize that they are in a safe place for experimentation. Help clients move from despair or indifference to hope.
- help clients become more mindful about their problems and opportunities. Mindfulness is the opposite of routinized conformity and passive learning.
- help clients break with self-restricting mind-sets. Help them move from an "I can't" to an "I can" mind-set.

However, once you know the conditions that favor creativity, you can use responding and challenging skills to help clients awaken whatever creative resources they might have.

Divergent Thinking

Many people habitually take a convergent-thinking approach to problem solving; that is, they look for *one* right answer. Such thinking has its uses; however, many of the problem situations of life are too complex to be handled by convergent thinking. Such thinking limits the ways in which people use their own and environmental resources.

In contrast, divergent thinking—thinking "outside the boxes"—assumes that there is always more than one answer. In the terms that concern us here, this means more than one way to manage a problem or develop an opportunity. Unfortunately, divergent thinking, as helpful as it can be, is not always rewarded in our culture and sometimes is even punished. For instance, students who think divergently can be thorns in the sides of teachers. Some teachers feel comfortable only when they ask questions in such a way as to elicit what they consider the one right answer. When students who think divergently give answers that are different from the ones expected, they may be ignored, corrected, or punished—even though their responses might be quite useful, perhaps even more useful than the expected responses. Students learn that divergent thinking is not rewarded, at least not in school. They may then generalize their experience and end up thinking that divergent thinking is simply not a useful form of behavior. Consider the following case.

> Quentin wanted to be a doctor, so he enrolled in the pre-med program at college. He did well, but not well enough to get into medical school. When he received the last notice of refusal, he said to himself: "Well, that's it for me and the world of medicine. Now what will I do?" When he graduated, he took a job in his brother-in-law's business. He became a manager and did fairly well financially, but he never experienced much career satisfaction. He was glad that his marriage was good and his home life rewarding, because he derived little satisfaction from his work.

Not much divergent thinking went into handling this problem situation. No one asked Quentin what he really wanted. For Quentin, becoming a doctor was the "one right career." He didn't give serious thought to any other career related to the field of medicine, even though there are dozens of interesting and challenging jobs in the allied health sciences.

The case of Asif, who also wanted to become a doctor but failed to get into medical school, is quite different from that of Quentin.

> Asif thought to himself: "Medicine still interests me; I'd like to do something in the health field." With the help of a medical career counselor, he reviewed the possibilities. Even though he was in pre-med, he had never realized that there were so many careers in the field of medicine. He decided to take whatever courses and practicum experiences he needed to become a nurse. Then, while working in a clinic in the hills of Appalachia, where he found the experience invaluable, he managed to get an M.A. in family-practice nursing by attending

a nearby state university part-time. He chose this specialty because he thought that it would enable him to be closely associated with delivery of a broad range of services to patients and would also enable him to have more responsibility for the delivery of these services.

When Asif graduated, he entered private practice with a doctor as a nurse practitioner in a small midwestern town. Since the doctor divided his time among three small clinics, Asif had a great deal of responsibility in the clinic where he practiced. He also taught a course in family-practice nursing at a nearby state school and conducted workshops in holistic approaches to preventive medical self-care. Still not satisfied, he began and finished a doctoral program in practical nursing. He taught at a state university and continued his practice. Needless to say, his persistence paid off with an extremely high degree of career satisfaction.

A successful professional career in health care remained Asif's aim throughout. A great deal of divergent thinking and creativity went into the elaboration of this aim into specific goals and into coming up with the courses of action to accomplish them. But for every success story like this, there are many more failures. Quentin's case is the norm, not Asif's. For many, divergent thinking is uncomfortable.

We now turn our attention to two skills or methodologies: brainstorming, which is a method for coming up with problem-managing goals and the strategies for accomplishing them, and the balance-sheet, which is a method for evaluating the usefulness of goals and the strategies for accomplishing them. The former helps clients create ideas. The latter helps them evaluate the ideas they create.

Brainstorming: A Tool for Divergent Thinking

One excellent way of helping clients think divergently and more creatively is brainstorming. Brainstorming is a simple idea-stimulation technique for exploring the elements of complex situations. As such, it could be used in Stage I to help clients explore what is problematic in their lives or the dimensions of a single problem. By using a series of open-ended questions and probes, one counselor helped a priest, who was talking very vaguely about the problems of his life, lay all of his cards on the table. Eventually the priest revealed that his dissatisfactions included doubts about the relevance of his profession, the loss of respect for the priesthood on the part of the laity, belonging to an organization that seemed to be going "downhill," sexual ambiguity and conflict, overwork, trivial demands on his time, a lack of volunteers in his parish, unrealistic expectations on the part of parishioners, disaffection toward superiors, impoverished social life, and mild depression that continually sapped his energy. In this case, brainstorming helped produce a range of issues with substance to them. Many clients, however, would find such a process overwhelming. When appropriate, brainstorming can be used as a tool in Stages II and III for helping clients develop possibilities for a better future and ways of accomplishing goals.

There are certain rules that help make this technique work: suspend judgment, produce as many ideas as possible, use one idea as a takeoff point for others, get rid of normal constraints to thinking, and produce even more ideas by clarifying items on the list. Helping clients brainstorm possibilities is a critical skill in Stage II, Stage III, and during the implementation phase of problem management. It is a skill that helpers should think of giving away to their clients as part of the helping process. Here, then, are the rules, in more detail.

Suspend your own judgment, and help clients suspend theirs. In the brainstorming phase, do not let clients criticize the ideas they are generating and do not criticize these possibilities yourself. There is some evidence that this rule is especially effective when the problem situation has been clarified and defined and goals have been set. In the following example, a man who is in pain because he has been rejected by a woman he loves is exploring ways of getting her out of his mind.

CLIENT: One possibility is that I could move to a different city, but that would mean that I would have to get a new job.
HELPER: Add it to the list. Remember, we'll discuss and critique them later.

Having clients suspend judgment is one way of handling some clients' tendency to play "Yes, but" games with themselves; that is, to come up with a good idea and then immediately show why it isn't really a good idea. By the same token, don't let yourself say such things as "Explain what you mean," "I like that idea," "This one is useful," "I'm not sure about that idea," or "How would that work?" Premature criticism cuts down on creativity. A marriage counselor was helping a couple brainstorm possibilities for a better future. When Nina said, "We will stop bringing up past hurts," Tip replied, "That's your major weapon when we fight. You'll never be able to give that up." The helper said, "Add it to the list. We'll look at the realism of these possibilities later on."

Encourage clients to come up with as many strategies as possible. The principle is that quantity ultimately breeds quality. Some of the best ideas come along later in the brainstorming process. Cutting the process short can be self-defeating.

CLIENT: Maybe that's enough. We can start putting it all together.
HELPER: It didn't sound like you were running out of ideas.
CLIENT: I'm not. It's actually fun. It's liberating.
HELPER: Well, let's keep on having fun for a while.

Within reason, the more ideas the better. Helping clients identify many possibilities for a better future increases the quality of the possibilities that are eventually chosen and turned into goals. I once worked with a client who brainstormed a few ideas and then wanted to stop. I pushed him. He generated a few more. Then another nudge. In the end, he generated over 25 strategies. About seven of the strategies were used in his plan. But five out of the seven came from the last dozen strategies brainstormed. If we had stopped, he would still have formulated an action plan — but it would not have been as good.

Help clients use one idea as a takeoff point for another. This is called piggybacking. Without criticizing the client's productivity, encourage the client to expand strategies already generated and to combine different ideas to form new possibilities. In the following example, the client is trying to come up with ways of increasing her self-esteem.

CLIENT: One way I can get a better appreciation of myself is to make a list of
       my accomplishments every week, even if this means just getting through
       the ordinary tasks of everyday life.
HELPER: Expand that a bit for me.
CLIENT (pausing):   Well, I could star the ones that took some kind of special ef-
       fort on my part and celebrate them with my husband.

Variations and new twists in strategies already identified are taken as new ideas. In group settings, which includes marriage counseling, clients are encouraged to piggyback on others' ideas without criticizing them. When one spouse said, "We will collaborate in drawing up the household budget," the other spouse said, "And the budget will include some 'mad money' for each of us."

A form of piggybacking is possible with individual clients. When one client with multiple sclerosis said, "I'll have a friend or two with whom I can share my frustrations as they build up," the helper asked, "What would that look like?" The client replied, "It would not be just a poor-me thing. It would be a give-and-take relationship. We'd be sharing both joys and pains like other people do."

Help clients let themselves go and develop some "wild" ideas. When clients seem to be drying up or when the strategies being generated are quite pedestrian, I often say, "Okay, now draw a line under the items on your list and write the word *wild* under the line. Now let's see if we can develop some really wild ways of getting your goals accomplished." Later it is easier to cut suggested strategies down to size than to expand them. The wildest possibilities often have within them at least a kernel of an idea that will work.

HELPER: So you need money for school. Well, what are some wild ways you
       could go about getting the money you need?
CLIENT: Well, let me think. . . . I could rob a bank . . . or print some of my own
       money . . . or put an ad in the paper and ask people to send me money.

This client ended up "printing money" by starting a modest desktop publishing business. Clients often need permission to let themselves go even in such a harmless way. Too often we repress good ideas because when they are first stated they sound foolish. The idea is to create an atmosphere in which such apparently foolish ideas will not only be accepted but encouraged.

When it comes to acting, help clients think of conservative strategies, liberal strategies, radical strategies, and even outrageous strategies.

Clients who are stuck are helped by including wild possibilities. Often there is a very realistic possibility hidden in the wildness. A client talking to a counselor in a sex-addiction clinic said, "I'll become a monk." Later on, the counselor asked, "What would a modern-day monk—who's not even a Catholic—look like?" This helped the client explore the concept of sexual responsibility from a completely different perspective and to rethink the place of religion in his life.

Help clients clarify items on the list. Without criticizing their proposals, help clients clarify them. When an idea is clarified, it can be expanded. In the following example, a college student is talking about getting financial support now that he has moved out of the house and his parents have cut off funds.

CLIENT: . . . And I suppose there might be the possibility of loans.

COUNSELOR: What kind of loans? From whom?

CLIENT: Well, my grandfather once talked about taking out a stake in my future, but at the time I didn't need it. (Pause.) And then there are the low-interest state loans. I never think of them because none of the people I know use them. I bet I'm eligible. I'm not even sure, but the school itself might have some loans.

Clarifying ideas should not be a way of stopping the process or justifying items already brainstormed. Done right, it is an avenue to further possibilities.

Brainstorming is not the same as free association. While clients are encouraged to think of even wild possibilities, still these possibilities must in some way be stimulated by and relate to the client's problem situation and chosen goals.

Balance-Sheet Methodology

Since choosing goals and strategies to accomplish them often requires choosing from among a number of alternatives, counselors need ways of helping clients do so. Decision making, as we saw in Chapter 10, is a messy process. Part of the messiness of decision making is the failure on the part of clients to consider the consequences of the decisions they make during the helping process. For instance, a counselor who helped the homeless found that some of her clients rejected out of hand the notion of any kind of shelter. They prized their freedom. Some of them, however, did not weigh the consequences of their decision and this counselor challenged them to do so. The balance sheet, illustrated in Figure 11-1, is a decision-making aid that counselors can use to help clients through difficult choices (*see* Janis, 1983; Janis & Mann, 1977; Wheeler & Janis, 1980). It is a way of helping clients examine the consequences of their choices in terms of utility—"Will this goal help me move beyond the problems that I am currently struggling

If I choose this course of action:		
The self		
Gains for self:	Acceptable to me because:	Not acceptable to me because:
Losses for self:	Acceptable to me because:	Not acceptable to me because:
Significant others		
Gains for significant others:	Acceptable to me because:	Not acceptable to me because:
Losses for significant others:	Acceptable to me because:	Not acceptable to me because:
Social setting		
Gains for social setting:	Acceptable to me because:	Not acceptable to me because:
Losses for social setting:	Acceptable to me because:	Not acceptable to me because:

FIGURE 11-1
The Decision Balance Sheet

with?'' — and acceptability — ''Can I live with this choice?'' Ambivalence — ''Should I or shouldn't I?'' — often paralyzes clients. The balance sheet helps clients move beyond ambivalence.

The balance sheet can be used to help clients choose goals and strategies for accomplishing goals (Hesketh, Shouksmith, & Kang, 1987). As you can readily see, the balance sheet is rather thorough. Therefore, it need never be used in its entirety nor should it be used to help clients evaluate every decision they make. As with other helping methods, clients should be helped to use it only when using it will add value. In subsequent chapters, we will see brief examples of how to use the balance sheet. The counselor just mentioned used it in bits and pieces in her attempts to help the homeless. For instance, she tried to help a homeless mother with a young daughter balance her loss of freedom in the shelter against the dangers her daughter might experience if they were to remain on the street. The counselor did not engage in scare tactics, but she did challenge her clients.

STAGE II VERSUS STAGE III: ACTIVITIES, OUTCOMES, AND IMPACT

The distinction between Stage II—helping clients determine what they want—and Stage III—helping clients discover the means they are going to use to get what they want—is a very important one. It is the distinction between *what* they want and *how* they are going to get what they want. It is emphasized here because, at least initially, clients and sometimes even helpers do not keep this important distinction in mind. When put in moral terms, the distinction can be shocking. For instance, if a client says that his broad goal is to achieve a more comfortable lifestyle and that the means he will use to achieve this lifestyle will include lying, cheating, and stealing, you might well be shocked. The goal seems reasonable, yet the means to be used to achieve the goal do not.

Furthermore, this distinction between goals and the means to achieve them is important for more than moral reasons. Clients, when helped to explore what is going wrong in their lives, often ask, ''Well, what should I do about it?'' They focus on solutions in terms of *activities*. But, as we shall see, essential as activities are, they are valuable only to the degree that they produce *outcomes*. An outcome is a goal, something a client wants or needs. Even more important than the achievement of a goal, however, is the *impact* that it has on the life of the client—that is, how it relates to managing problem situations and developing unused opportunities.

If a client with marital problems says that she wants to get away from her husband for a while, she is talking about an activity rather than a goal. The helper might well probe to see what this mini-separation is meant to accomplish and what impact it might have. If the client replies, ''Well, there has been so much tension between the two of us lately, we have not been able to talk decently to each other. We yell. We pout, We withdraw. We fight.

We both think some kind of moratorium will help." Now goals and the impact these goals will have on their lives begin to emerge — reduced stress, a calmer atmosphere that promotes dialogue rather than fighting, an opportunity for each to get his and her thoughts together, and so forth are desirable outcomes. The potential impact of these outcomes is quite positive. In a calmer atmosphere they can begin working out their problems and determine how they want to reinvent their marriage.

The distinction between activities, outcomes, and impact is also seen in the following example.

> Enid, a 40-year-old single woman, is making a great deal of progress in controlling her drinking through her involvement with an AA program. She engages in certain *activities*; for instance, she attends AA meetings, follows the 12 steps, stays away from situations that would tempt her to drink, and calls fellow AA members when she feels depressed or when the temptation to drink is pushing her hard. The *outcome* is that she has been sober for over seven months. She feels that this is a significant accomplishment. The *impact* of all this is quite rewarding. She feels better about herself and she has had the energy and enthusiasm to do things that she has not done in years — developing a circle of friends, getting interested in church activities, and doing a bit of travel.
>
> But Enid is also struggling with a troubled relationship with a man. In fact, her drinking was, in part, an ineffective way of avoiding the problems in the relationship. She knows that she no longer wants to tolerate the psychological abuse she has been getting from her male friend, but she fears the vacuum she will create by cutting off the relationship. She is, therefore, trying to determine what she wants, realizing that ending the relationship might turn out to be the best option.
>
> She has engaged in a number of *activities*. For instance, she has become much more assertive in the relationship. She now cuts off contact whenever her companion becomes abusive. She does not let him make all the decisions in the relationship. But the relationship remains troubled. Even though she is doing a lot of things, there is no satisfactory *outcome*. She has not yet determined what the outcome should be; that is, she has not determined what kind of relationship she would like and if it is possible to have such a relationship with this man. Nor has she determined to end the relationship.
>
> Finally, after one seriously abusive episode, she tells him that she is ending the relationship. She does what she has to do to sever all ties with him (*activities*) and the *outcome* is that the relationship ends and stays ended. The *impact* is that she feels lonely but liberated. She begins expanding her social life through her church contacts. She joins a self-help group of single women like herself in order to find out whether she wants an intimate relationship with a man — and, if so, what to do about it.

In Stage II, clients are helped to determine what they need and want — *outcomes*. In Stage III, they are helped to figure out what they need to do to get what they want — *activities*. However, skilled counselors continually help clients get a clear idea of the *impact* of the activities they engage in and the goals they are determined to pursue. In one sense, helping is all about impact.

HELPING CLIENTS CREATE A BETTER FUTURE

Stage II focuses on a better future, the client's preferred scenario. Problems can make clients feel hemmed in and closed off. To a greater or lesser extent they have no future, or the future they have looks troubled. As Gelatt (1989) noted, "The future does not exist and cannot be predicted. It must be imagined and invented" (p. 255). The steps of Stage II outline three ways in which helpers can be with their clients at the service of exploring and developing this better future.

Step II-A. Helping clients identify *possibilities* for a better future.

Step II-B. Helping clients choose specific preferred-scenario *goals*.

Step II-C. Helping clients discover incentives for *commitment* to their goals.

Without minimizing in any way what helpers can help their clients accomplish through Stage-I interventions—that is, problem and opportunity clarification; the development of new, more constructive perspectives of self, others, and the world; the choice of high-leverage issues to work on—the real power of helping clients *do something* about the problem situations and unused opportunities comes about in Stage II. O'Hanlon and Weiner-Davis (1989) claimed that a trend "away from explanations, problems, and pathology, and toward solutions, competence, and capabilities" (p. 6) was emerging in the helping professions. Although I would hope that this is (as they suggested) a "megatrend," I am not as convinced. One study showed that clients are more interested in solutions to their problems and feeling better, while helpers are more concerned about the origin of problems and transforming them through insight (Llewelyn, 1988). Furthermore, the tendency persists to equip helpers-to-be with the skills needed to help clients understand their problems but to stop short of providing them with the skills needed to help clients move to problem-managing action.

There is a difference between working through problems and transcending or moving beyond them. Working through them involves exploring a problem or unused opportunity from many different angles. Clients come to understand the roots of their problems. The transcending approach suggests that examining problem situations is, in many cases, of limited value. Problems are not managed by merely understanding them. Opportunities are not developed by seeing them clearly. In the end there is no formula. There must be some kind of balance between understanding and acting and this balance will look different for each client or even the same client in different situations. But the goal of helping, as stated in Chapter 1, is "problems managed," not "problems understood." Understanding is a subgoal—often a critical subgoal—on the way to the principal goal of managing.

The three steps of Stage II become clearer to clients when they are translated into everyday language. The questions that follow are asked, of course, against the background of some degree of understanding on the part of the client of the problem situation or of the unused opportunity.

Step II-A.   What do you want? What do you need? What are some of the possibilities?

Step II-B.   Given the possibilities, what do you really want? What are your choices?

Step II-C.   What are you willing to pay for what you want?

Helping clients answer these questions increases the probability of their acting on their own behalf. Finally, even though these questions are treated separately for the sake of learning different helping methods, in practice they are often intermingled.

STEP II-A: A BETTER FUTURE — WHAT DO YOU WANT? POSSIBILITIES

THE GOAL OF STEP II-A

HELPING CLIENTS DISCOVER WHAT THEY WANT AND NEED

BRAINSTORMING TO DISCOVER CLIENTS' WANTS AND NEEDS

Future-Oriented Questions and Probes

Helping Clients Find Models

A FUTURE-CENTERED APPROACH TO HELPING

EVALUATION QUESTIONS FOR STEP II-A

THE GOAL OF STEP II-A

The goal of Step II-A, illustrated in Figure 12-1, is to help clients develop a sense of direction by exploring possibilities for a better future. I once was sitting at the counter of a late-night diner when a young man sat down next to me. The conversation drifted to the problems he was having with a friend of his. I listened for a while and then asked, "Well, if your relationship was just what you wanted it to be, what would it look like?" It took him a bit to get started, but eventually he drew a picture of the kind of relationship he could live with. Then he stopped, looked at me, and said, "You must be a professional." I believe he thought that because this was the first time in his life that anyone had ever asked him to describe the possibilities of a better future.

The power of possibilities.   Too often, the exploration and clarification of problem situations are followed, almost immediately, by the search for solutions. *Solutions* is an ambiguous word. It can refer to the series of actions that will lead to the resolution of the problem situation or the development of an opportunity. But it can also refer to what will be *in place* once these actions are completed. There is great power in visualizing outcomes, just as there is a danger in formulating action strategies before getting a clear idea of desired outcomes. Stage II is about results, outcomes, and accomplishments. Stage III is about strategies and plans for delivering these outcomes.

Possible selves.   Clients come to helpers because they are stuck. Counseling is a process of helping clients get unstuck and develop a sense of direction. Consider the case of Ernesto. He was very young, but very stuck for a variety of socio-cultural and emotional reasons.

> A counselor first met Ernesto in the emergency room of a large urban hospital. He was throwing up blood into a pan. He was a street gang member, and this was the third time he had been beaten up in the last year. He had been so severely beaten this time that it was likely that he would suffer permanent physical damage. Ernesto's style of life was doing him in, but it was the only one he knew. He was in need of a new way of living, a new scenario, a new way of participating in city life. This time he was hurting enough to consider the possibility of some kind of change.

Markus and Nurius (1986) used the term *possible selves* to represent "individuals' ideas of what they might become, what they would like to become, and what they are afraid of becoming" (p. 954). The counselor worked with Ernesto, not by helping him explore the complex socio-cultural and emotional reasons why he was in this fix, but principally by helping him explore his possible selves in order to discover a different purpose in life, a different direction, a different lifestyle. Step II-A is about possible selves.

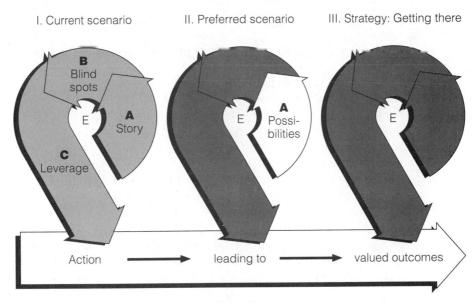

FIGURE 12-1
Step II-A: Developing Preferred-Scenario Possibilities

HELPING CLIENTS DISCOVER WHAT THEY WANT AND NEED

Step II-A can be seen as a way of helping clients develop a wish list for a better future. What do you want? What would life look like if it looked better? What would this problem situation look like if it were being managed? What would this unused opportunity look like if it were being developed?

What do you want?   Tom was HIV-positive and it was clear that he was getting sicker. But he wanted to "get some things done" before he died. Tom's action orientation helped a great deal. Over the course of a few months, a counselor helped him to name some of the things he wanted before he died or on his journey toward death. Over time, Tom came up with the following possibilities:

- He wanted someone with a religious orientation, such as a minister, with whom he could occasionally talk about the bigger issues of life and death.
- He wanted to die with a sense of meaning.
- He wanted to be a member of some kind of community, maybe a self-help group of fellow AIDS victims, people who did not fear him.
- He wanted fewer financial worries.

- He wanted one or two intimates with whom he could share the ups and downs of daily life.
- He wanted to be engaged in some kind of productive work, whether paid or not, as long as he could.
- He wanted a decent place to live, maybe with others.
- He wanted access to decent medical attention from medical staff who would not treat him like a new-age leper.
- He wanted to manage bouts of anxiety and depression better.
- He wanted to be reintegrated with his family.
- He wanted especially some kind of reconciliation with his father.
- He wanted to make peace with one or two of his closest friends who abandoned him when they learned he was HIV-positive.
- He wanted to die in his hometown.

The counselor helped Tom stitch together a set of goals from these possibilities (Stage II) and ways of accomplishing them (Stage III). This help contributed substantially to Tom's quality of life even under very difficult circumstances.

What do you need? What clients want is not always the same as what they need. Wants and needs must be placed in balance.

> Irv, a 41-year-old entrepreneur, collapsed one day at work. He had not had a physical in years. He was shocked to learn that he had both a mild heart condition and multiple sclerosis. His future was uncertain. The father of one of his wife's friends had multiple sclerosis but had lived and worked well into his 70s. But no one knew what the course of Irv's disease would be. Since he had made his living by developing and then selling small businesses, he wanted to continue to do this. But this was too physically demanding. Working 60-70 hours per week, even though he loved it, was no longer in the cards. What he needed was a less physically demanding work schedule. Furthermore, he had always plowed the money he received from selling one business into starting up another. But now he needed to think of the future financial well-being of his wife and three children. Up to this point his philosophy had been that the future would take care of itself. It was very wrenching for him to move from a lifestyle he wanted to one he needed.

Involuntary clients often need to be challenged to look beyond their wants to their needs. One woman who voluntarily led a homeless life was attacked and severely beaten on the street. She still wanted the freedom that came with her lifestyle. A counselor suggested that there were ways of keeping her freedom even if she were to adopt a different lifestyle. When challenged to consider the kinds of freedom she wanted, she admitted that freedom from responsibility was at the core. It was her choice to live the way she wanted, but the counselor helped her explore the consequences of her choices.

What do you want and need? An individual case.   Probably most cases deal with both wants and needs.

> Milos had come to the U.S. as a political refugee. The last few months in his native land had been terrifying. He had been jailed and beaten and got out just before another crackdown. Once the initial euphoria of having escaped had subsided, he spent months feeling confused and disorganized. He tried to live as he had in his own country, but the culture in the U.S. was too invasive. He thought he should feel grateful and yet he felt hostile. After two years of misery, he began seeing a counselor. He had resisted getting help since "back home" he had been "his own man."
>
> In discussing these issues with a counselor, it gradually dawned on Milos that he *wanted* to re-establish links with his native land but that he *needed* to integrate himself into the life of his host country. He saw that the accomplishment of this twofold goal would be very freeing. He began finding out how other immigrants who had been in the U.S. longer than he had accomplished this goal. He spent time in the immigrant community rather than the refugee community. In the immigrant community there was a long history of keeping links to the homeland culture alive. But the immigrants had also adapted to the best in their adopted country. The friends he made became role models for him. The more active he became there, the more his depression lifted.

Had Milos focused only on what he wanted or what he needed, he would have remained unhappy.

What do you want and need? A family case.   This case is more complex because it involves a family. Not only does the family as a unit have its wants and needs, but each of the individual members has his or her own. Therefore, it is even more imperative to review possibilities for a better future so that competing needs can be reconciled.

> Lane, the 15-year-old son of Troy and Rhonda Washington, was hospitalized with what was diagnosed as an acute schizophrenic attack. He had two older brothers, both teenagers, and two younger sisters, one 10 and one 12, all living at home. The Washingtons lived in a large city. Although both parents worked, their combined income still left them pinching pennies. They also ran into a host of problems associated with their son's hospitalization—the need to arrange ongoing help and care for Lane, financial burdens, behavioral problems among the other siblings, and marital conflict. Further, they felt a social stigma in the community: "They're a funny family with a crazy son." or "What kind of parents are they?" To make things worse, they did not think the psychiatrist and psychologist they met at the hospital took the time to understand their concerns. They felt that the helpers were trying to push Lane back out into the community. In their eyes, the hospital was "trying to get rid of" Lane; the family complained that the hospital gave Lane "some pills and then gave him back." No one explained to them that short-term hospitalization was meant to guard the civil rights of patients and avoid the negative effects of longer-term institutionalization.

When Lane was discharged, his parents were told that he might have a relapse, but they were not told what to do about it. They faced the prospect of caring for Lane in a climate of stigma without adequate information, services, or relief. Feeling abandoned, they were very angry with the mental-health establishment. They had no idea what they should do to respond to Lane's illness or to the range of family problems that had been precipitated by the episode. By chance, the Washingtons met someone who had worked for the National Alliance for the Mentally Ill (NAMI), an advocacy and education organization. This person referred them to an agency that provided support and help.

What does the future hold for such a family? With help, what kind of future can be fashioned? Social workers at the agency helped the Washingtons identify needs and wants in seven areas (*see* Bernheim, 1989).

1. *The home environment.*   The Washingtons wanted/needed an environment in which the needs of all the family members were balanced. They didn't want their home be an extension of the hospital. They wanted Lane taken care of, but they wanted to attend to the needs of the other children and to their own needs as well.

2. *Care outside the home.*   They wanted a comprehensive therapeutic program for Lane. They needed to review possible services, identify relevant services, and arrange access to these services. They needed to find a way of paying for all this.

3. *Care inside the home.*   They wanted all family members to know how to cope with Lane's residual symptoms. He might be withdrawn or aggressive, but they needed to know how to relate to him and help him handle behavioral problems.

4. *Prevention.*   Family members needed to be able to spot early warning symptoms of impending relapse. They also needed to know what to do when they saw such signs, including such things as contacting the clinic or, in the case of more severe problems, arranging for an ambulance or getting help from the police.

5. *Family stress.*   They needed to know how to cope with the increased stress all of this would entail. They needed forums for working out their problems. They wanted to avoid family blow-ups and when blow-ups occurred, they wanted to manage them without damage to the fabric of the family.

6. *Stigma.*   They wanted to understand and be able to cope with whatever stigma might be attached to Lane's illness. For instance, the children needed to know what to do and what not to do if or when they were taunted for having a "crazy" brother. Family members needed to know whom to tell, what to say, how to respond to inquiries, and how to deal with blame and insults.

7. *Limitation of grief.*   They needed to know how to manage the normal guilt, anger, frustration, fear, and grief that go with problem situations like this.

Bernhein's schema constituted a useful checklist for stimulating thinking about possibilities for a better future. The Washingtons first needed to be helped to develop these possibilities. The next step would be to help them set priorities and establish goals to be accomplished.

BRAINSTORMING TO DISCOVER CLIENTS' WANTS AND NEEDS

Since clients mired down in problem situations don't always know upfront what they want and need, counselors add value by helping them brainstorm and explore possibilities. This is especially true with clients who are not future-oriented, clients with restricted imaginations, clients rooted in the past or the misery of the present, and clients with poor conceptual and verbal skills. Helpers should not expect too much or too little from their clients. They are expecting too much if they believe that most clients, on their own, can come up with a broad range of preferred-scenario possibilities and then compose these into a neatly painted picture of a better future. On the other hand, helpers are expecting too little of most clients if they see them as totally unimaginative and uncreative.

All the communications skills discussed earlier—attending, listening, understanding, empathy, probing, and challenging at the service of creativity, divergent thinking, and brainstorming—are needed in Stages II and III. These skills are essential for using methods to help clients discover what they want. Two methods are outlined here: using future-oriented questions and probes and helping clients find models.

Future-Oriented Questions and Probes

One way of helping clients invent the future is to ask them, or get them to ask themselves, future-oriented questions related to their current unmanaged problems or undeveloped opportunities. The questions given here are different ways of helping clients answer the questions, "What do you want? What do you need?" They all deal with wants and needs in terms of what will be *in place* after they act. These questions are often more productively pursued if you also use the brainstorming guidelines discussed in Chapter 11.

• *What would this problem situation look like if you were managing it better?* Ken, a college student who has been a loner, has been talking about his general dissatisfaction with his life. In response to this question, he said: "I'd be having fewer anxiety attacks. And I'd be spending more time with people rather than by myself."

• *What changes in your present lifestyle would make sense?* Cindy, who described herself as a "bored homemaker," replied, "I would not be drinking as much. I'd be getting more exercise. I would not sit around and watch the soaps all day. I'd have something meaningful to do."

- *What would you be doing differently with the people in your life?* Lon, a graduate student at a university near his parents' home, realized that he had not yet developed the kind of autonomy suited to his age. He mentioned these possibilities: "I would not be letting my mother make my decisions for me. I'd be sharing an apartment with one or two friends."

- *What patterns of behavior would be in place that are not currently in place?* Bridget, a depressed resident in a nursing home, said, "I'd be engaging in more of the activities offered here in the nursing home."

- *What current patterns of behavior would be eliminated?* Bridget added, "I would not be putting myself down for physical ailments I cannot control. I would not be complaining all the time. It gets me and everyone else down!"

- *What would you have that you don't have now?* Sissy, a single woman who has lived in a housing project for 11 years, said, "I'd have a place to live that's not rat-infested. I'd have some friends. I wouldn't be so miserable all the time."

- *What accomplishments would be in place that are not in place now?* Ryan, a divorced man in his mid-30s, said, "I'd have my degree in practical nursing. I'd be doing some part-time teaching. I'd have someone to marry."

- *What would this opportunity look like if you developed it?* Sophie, a woman with a great deal of talent but who feels like a second-class citizen in the company in which she works, had this to say: "In two years I'll be an officer of this company or have a very good job in another firm."

Of course, these are not the only questions that can prompt the exploration of possibilities for a better future. Effective helpers learn how to formulate questions and probes that respond to each client's needs.

It is a mistake to suppose that clients will automatically gush with answers. Ask the kinds of questions just listed, or encourage clients to ask themselves the questions, but then help them answer the questions. Many clients don't know how to use their innate creativity. Thinking divergently is not part of their mental lifestyle. You have to work with clients to help them produce some creative output. Some clients are reluctant to name possibilities for a better future because, as noted in Chapter 11, they sense that this will bring more responsibility. They will have to move into action mode.

Probes can also be stated as sentence stems to be completed by the client. This is another approach to brainstorming.

- Here's what I need . . .
- Here's what I want . . .
- Here are some items on my wish list . . .
- When I'm finished I will have . . .
- There will be . . .
- I will have in place . . .

- I will consistently be . . .
- There will be more of . . .
- There will be less of . . .

The purpose of the probes and sentence stems is to help clients be more creative, to engage in divergent thinking, to move outside the walls they may have created around themselves.

Helping Clients Find Models

Some clients can see future possibilities better when they see them embodied in others. You can help clients brainstorm possibilities for a better future by helping them to identify models. By models I don't mean superstars or people who do things perfectly. This would be self-defeating. In the next example, a marriage counselor is talking with a middle-aged, childless couple. They are bored with their marriage. When the counselor asked them, "What would your marriage look like if it looked a little better?" he could see that they were stuck.

COUNSELOR: Maybe the question would be easier to answer if you reviewed some of your married relatives, friends, or acquaintances.
WIFE: None of them have super marriages. (Husband nods in agreement.)
COUNSELOR: No, I don't mean super marriages. I'm looking for things you could put in your marriage that would make it a little better.
WIFE: Well, Fred and Lisa are not like us. They don't always have to be doing everything together.
HUSBAND: Who says we have to be doing everything together? I thought that was your idea."
WIFE: Well, we always *are* together. If we weren't always together, we wouldn't be in each other's hair all the time.

Even though it was a somewhat torturous process, these two people were able to come up with a range of possibilities for a better marriage. The counselor had them write these possibilities down so they wouldn't lose them. At this point the purpose was not to get the clients to commit themselves to these possibilities but to review them.

In the following case, the client finds herself making discoveries by observing people she had not identified as models at all.

Fran, a somewhat withdrawn college junior, realizes that when it comes to interpersonal competence, she is not ready for the business world she intends to enter when she graduates. She wants to do something about her interpersonal style and a few nagging personal problems. She sees a counselor in the Office of Student Services. After a couple of discussions with him, she joins a "life-style" group on campus that includes some training in interpersonal skills.

Even though she expands her horizons a bit from what the members of the group say about their experiences, behaviors, and feelings, she tells her counselor that she learns even more by watching her fellow group members in action. She sees behaviors that she would like to incorporate in her own style. A number of times she says to herself in the group: "Ah, there's something I never thought of." Without becoming a slavish imitator, she begins to incorporate into her own style some of the patterns she sees in others.

Models or exemplars can help clients name what they want more specifically. Models can be found anywhere: among the client's relatives, friends, and associates; in books; on television; in history; in movies. Counselors can help clients identify models, choose those dimensions of others that are relevant, and translate what they see into realistic possibilities for themselves.

There are other methods for helping clients discover what they want. Counselors can help clients review better times — that is, times when they had what they wanted. Married couples, substance abusers, and others can benefit from reviewing better times — the best in the past can point toward a better future. Some counselors use writing approaches to help their clients determine what they want. One helper had his clients write the kinds of résumés they would like to be able to have. Another helper had his clients write their epitaphs to consider the kinds of lives they would like to lead. Finally, some counselors help their clients set up experiments in living to help them open up new possibilities. One very self-centered client spent two weeks delivering meals to AIDS patients. He learned that he had to "get out of himself" if he was ever going to be happy. Skilled helpers can find endless ways of helping clients identify what they want — possibilities for a better future.

A Future-Centered Approach To Helping

The focus of this chapter is on a possibility-rich future rather than on a past that cannot be changed or a present that is problematic. What follows is a nonlinear, future-oriented version of the helping model.

1. Telling the story. First, help clients tell their stories. But it may be a mistake to get the client to reveal all the details of the problem situation all at once at the beginning of the helping encounter. Sometimes a relatively quick look at the highlights of the story is enough to begin with.

Jay, a single man, aged 25, tells a story of driving while drinking and of an auto accident in which he was responsible for the death of a friend as well as his own severe back injuries. He has told the story over and over again to many different people, as if telling it enough will make what happened go away. When Jay met the counselor he had already told his story a number of times to different people. Retelling the whole thing in great detail was not going to help. A summary was enough.

2. Possibilities for a better future: A sense of hope.   Help clients take an initial cut at exploring what they want. Describing what they want — their preferred-scenario possibilities — can help clients develop a sense of hope. They identify a better future early in the helping process, and this future pervades the rest of the process.

Jay outlines a range of possibilities. He outlines a future in which he is more responsible for his actions, he is not hounded by regret and guilt, he has made peace with the family of his friend, he is physically rehabilitated within the limits of his injuries, and he is coping with the physical and psychological consequences of the accident. This is not the time to set clear-cut behavioral goals. Rather the client is being helped to outline a picture, not draft a blueprint. The purpose of this quick move to the preferred scenario is to help the client focus on a better future rather than a problematic and unchangeable past.

3. Return to Stage I.   Much of the work of Stage I can be done more effectively in the light of possibilities for a better future.

- It is easier for clients to spot the real issues and to develop them in more detail. They also see more clearly the gaps between where they are and where they want to be.
- When clients outline what they really want, it is easier for them to identify and manage blind spots and develop new perspectives.
- When clients begin to sketch possibilities for a better future, it is easier for them to see which problems and which unused opportunities have leverage for them. They know better what to work on.

Even a brief visit to Stage II — what things would look like if the problem situation were being managed a bit better — can act like a spotlight that illuminates the client's story and gives it a different cast. The current scenario is now seen as a gap or a set of gaps. A gap is not the same as a problem. A gap implies the need for action to bridge the gap.

In our example, Jay is helped to explore some of the gaps between his current and preferred scenarios and to develop new perspectives. For instance, he comes to see his overpowering regret and guilt as partially self-inflicted. When he engages in self-recrimination, he is not in the right frame of mind either to make peace with the family of his friend or to get seriously involved in a physical rehabilitation program. Managing the current flood of self-defeating emotions is a point of leverage. Until now he has not seen this constant self-chastisement as actually standing in the way of what he wants. He has unwittingly decided that self-chastisement is "good" and that more is even better. At this point, the exploration of the problem situation is no longer an end in itself but a tool to be used to fashion the future.

4. Return to Stage II.   Help clients use what they have learned in their return to Stage II to complete the process of determining what they want and what they are willing to do to get what they want.

Through a dialogue between what is and what could be, Jay begins to formulate a set of goals for himself. He cannot change the past, but he can learn from it. The energy he has been putting into self-recrimination can be put into self-reformation. He begins to explore more specifically what a self-responsible Jay would look like: not a person hounded by guilt, but not the irresponsible "free spirit" that lives on the edge of disaster, either. He sees a person whose energy is not squashed, but channeled. These ideas begin to get translated into specific goals; for instance, the goals of the physical rehabilitation program in which he is involved.

5. Moving on to Stage III: Strategies, plans, and actions.   Help clients formulate and implement strategies and plans to accomplish goals. As part of this process, Jay is helped to develop and implement strategies for controlling the flood of emotions he has let paralyze him. He uses his guilt as a stimulus to develop a way of meeting and making peace with the family of his friend. As he develops new perspectives, he begins to formulate strategies for making his emotions his allies instead of his enemies.

The purpose of this brief example is to illustrate a different, more upbeat movement in the helping process, what might be called a Stage II-centered approach to helping. It demonstrates that the helping process, at its best, is not a slavish, linear working through of the three stages.

Listen to a marriage counselor friend of mine who has adopted this back-and-forth approach in his work with couples.

> When a couple comes, they don't need to spend a great deal of time talking about what is going wrong. They know what's going wrong because they're living it every day. They are only too willing to tell their stories in great detail, but they tell entirely different stories. It's as if they were not even married to each other. When telling their stories, they begin to fight the way they fight at home. I don't find this very useful. And so, early on I make them focus on the future. I ask them questions like, "What do you want, individually and together? What would your marriage look like if it were a little bit better, not perfect, but just a little bit better?" I have them brainstorm separately and then share the possibilities. Instead of fighting, they are actually working together. If they get stuck, I ask them to think about other couples they know, couples that seem to have decent, if not perfect, marriages. "What's their marriage like? What parts of it would you like to see incorporated into your own mar-

riage?'' What I try to do is to help them use their imaginations to create a vision of a marriage that is different from the one they are experiencing. It's not that I never let them talk about the present. Rather, I try to help them talk about the present by having them engage in a dialogue between a better future and the unacceptable present.

This is not intended as a quick lesson in marriage counseling, but to provide the flavor of a more future-oriented approach.

- How at home am I in working with my own imagination?
- In what ways can I apply the concept of possible selves to myself?
- What problems do I experience as I try to help clients use their imaginations?
- Against the background of problem situations and unused opportunities, how well do I help clients focus on what they want?
- How effectively do I use empathy, a variety of probes, and challenge to help clients brainstorm what they want?
- Besides direct questions and other probes, what kinds of strategies do I use to help clients brainstorm what they want?
- How effectively do I help clients identify models and exemplars that can help them clarify what they want?
- How easily do I move back and forth in the helping model, especially in establishing a dialogue between Stage I and Stage II?
- How well do I help clients act on what they are learning?

STEPS II-B AND II-C: CHOICES AND COMMITMENT

Step II-B involves clients' translating possibilities into choices in terms of a set of goals. Step II-C deals with the client's initial commitment to these goals. Because Step II-C deals with initial commitment to a change, it is not a step in the strict sense. It is given special consideration, however, because part of the shadow side of helping is the fact that many clients set goals and develop programs for constructive change and then do little about them. Setting goals is cheap; doing something about them can be costly.

Step II-B: What Do Clients Really Want?

Once possibilities for a better future have been developed, clients need to make some choices. They need to choose one or more of these possibilities and turn them into a *program for constructive change* — a goal or a set of goals chosen to be accomplished. Step II-A is, in many ways, about *creativity*, getting rid of boundaries, thinking beyond a limited horizon, moving outside the box. Step II-B is about *innovation* — that is, creativity that has a positive impact on clients' lives. In this step, added in Figure 13-1, clients are helped to set practical goals, outcomes that have a problem-managing or opportunity-developing impact.

Moving from Possibilities to Choices

As noted in Chapter 11, a program for constructive change includes goals, together with the action strategies to be used to accomplish them. The first step in such a program is to outline possibilities for a better future (Step II-A). The second step is to choose specific goals (what I really want) from among these possibilities and give them shape (Step II-B). Consider the following case.

> Bea, a black woman, had been arrested when she went on a rampage in a bank and broke several windows. She was angry because she felt that she had been denied a loan mainly because she was black and a single mother. In discussing this incident with her minister, she came to see that she had a great deal of pent-up anger. She realized that venting her anger as she had done in the bank led to a range of negative consequences. She also realized that she still harbored a great deal of anger about social injustice and that she was taking it out on those around her, including her friends and two children. The minister helped her look at three possibilities: venting her anger, repressing her anger, and channeling her anger. She realized that the first was dysfunctional and that the second was demeaning. The third needed to be explored. In the end, Bea joined a political activist group involved in community organizing. She soon came to see that the politics of racial injustice were not relegated to election time. She learned that she could channel her anger without giving up her values or her intensity.

I. Current scenario   II. Preferred scenario   III. Strategy: Getting there

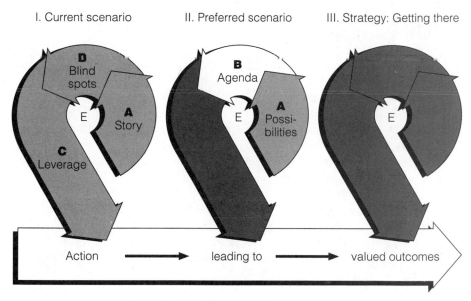

FIGURE 13-1
Step II-B: Crafting Productive Agendas

Once the immediate crisis had been handled, Bea had time to consider her choices. She moved to wanting to do "something" — that is, to react against what she had experienced — to wanting something much more substantial. The minister had helped her review her options and then shape the goal she finally settled on. The outcome was that Bea became an active participant in a political-action group. The impact on her life was quite positive. Not only did she channel her anger, but she also discovered that she was good at influencing others and getting things done. She felt very good about herself.

Methods for Helping Clients Shape Their Goals

Counselors can add value by using the communication skills and helping methods and techniques discussed up to this point to help clients choose, craft, shape, and develop their goals. Goals are specific statements about what clients want and need. The goals that emerge or that are explicitly set by clients in fashioning problem-managing and opportunity-developing programs for constructive change should have, at least eventually, certain characteristics in order to be *viable* goals. They should be

1. outcomes rather than activities,
2. specific enough to be verifiable and to drive action,
3. challenging and substantive,
4. realistic and sustainable,

5. in keeping with the client's values, and
6. set in a reasonable time frame.

In keeping with our no-formula approach, just how this package of goal characteristics will look in practice will differ from client to client. From a practical point of view, these characteristics can be seen as tools that counselors can use to help clients shape their goals. In this step, counselors help clients avoid the shadow-side pitfalls of decision making mentioned in Chapter 11. First, they help clients challenge themselves to make decisions, to name what they really want. Second, they help clients make decisions that have both substance and realism.

Help clients state what they want and need as outcomes or accomplishments.   The goal of counseling, as emphasized again and again, is neither discussing nor planning nor engaging in activities but problem-managing *results*. "I want to start doing some exercise" is an activity rather than an outcome. "Within six months I will be running three miles in less than 30 minutes at least four times a week" is an outcome, a pattern of behavior in place. If a client says, "My goal is to get some training in interpersonal communication skills," then she is stating her goal as a set of activities rather than as an accomplishment. If she says that she wants to become a better listener as a wife and mother, then she is stating her goal as an accomplishment.

You can help clients develop this past-participle approach — drinking stopped, number of marital fights decreased, anger habitually controlled — to describing what they want and need. Stating goals as outcomes or accomplishments is not just a question of language. Helping clients state goals as accomplishments rather than programs or activities helps them avoid directionless and imprudent action. If a client with AIDS says that he thinks he should join a self-help group, he should be helped to see what he wants to get out of such a group. Joining a group and participating in it are activities. Clients who know where they are going are more likely to work, not just hard, but smart.

Consider the case of Dillard, an ex-Marine diagnosed as suffering from post-traumatic stress syndrome.

> Dillard was involved in Desert Storm. Three of his buddies were killed by friendly fire. Afterwards, he began acting in strange ways, wandering around at times in a daze. He was sent home and given a medical discharge. Although he seemed to recover, he lived a rather aimless life. He went to college but dropped out during the first semester. He became rather reclusive but, as a friend who referred him to the helper said, "never really showed odd behavior." He moved in and out of a number of low-paying jobs. He also became less careful about his person. "You know, I used to be very careful about the way I dressed. Kind of proud of myself in the Marine tradition. Don't get me wrong; I'm not a bum and don't smell or anything, but I'm not myself." The whole direction of Dillard's life was wrong, and serious trouble could lie down the road. He was bothered by thoughts about what happened in Iraq and had taken to sleeping whenever he felt like it, day or night, "just to make it all stop."

Ed, Dillard's counselor, had a good relationship with Dillard. He helped Dillard tell his story and challenged some of his self-defeating thinking. He went on to help Dillard focus on what he wanted from life. They moved back and forth between Stage I and Stage II, between problems and possibilities for a better future. Eventually, Dillard began talking about his real needs and wants — what he needed to accomplish in order to get back to his old self.

DILLARD: I've got to stop hiding in my hole. I'm going to get out and see people more. I'm going to stop feeling so damn sorry for myself. Who wants to be with a nothing!

COUNSELOR: What will Dillard's life look like a year or two from now?

DILLARD: One thing for sure. He will be seeing women again. He might not be married yet but he will probably have a special girlfriend. And she will see him as an ordinary guy.

While Dillard talks about lifestyle changes in terms of patterns of behavior that will be in place, he is painting a picture of what he wants to be. The helper's probe reinforces this outcome approach. At this point it helps Dillard to talk about himself in the third person.

Help clients move from broad to clear and specific wants and needs. Stating wants and needs, not just as outcomes, but as *specific* outcomes, tends to motivate or drive behavior. Specific goals will, of course, be different for each client. Broad goals need to be translated into more specific goals and tailored to the needs and abilities of each client. Goal shaping includes helping clients develop the level of goal clarity and specificity they need in order to move into action. Skilled helpers use probes and challenges to help clients say what they really want to accomplish.

Dillard said that he wanted to become "more disciplined." His counselor helped him make this more specific.

COUNSELOR: Any particular areas you want to focus on?

DILLARD: Well, yeah. If I'm going to put more order in my life, I need to look at the times I sleep. I've been going to bed whenever I feel like it and getting up whenever I feel like it. It was the only way I could get rid of those thoughts and the anxiety. But I'm not nearly as anxious as I used to be. Things are calming down.

COUNSELOR: So more disciplined means a more regular sleep schedule because there's no particular reason now for not having one.

DILLARD: Yeah, now sleeping whenever I want is just a bad habit.

Dillard goes on to translate "more disciplined" into other problem-managing wants and needs related to school and work. Greater discipline will have a decidedly positive impact on his life. It will make him a doer rather than an onlooker.

Degrees of clarity and specificity in stating wants. Clients can be helped whenever necessary to move from vague desires to improve to quite specific goals. The movement might be from good intentions through broad goals to quite specific goals.

- *Good intentions.* "I need to do something about this" is a statement of intent. Some clients who say that they are going to do something about their lives then go out and accomplish something that helps them manage their problems or develop their opportunities better; however, other clients who say they are going to do something then do nothing. Some clients need a clearer picture of a better future before they act. In the following example, the client, Jon, has been discussing his relationship with his wife and children. The counselor has been helping him see how some of his behavior is perceived negatively by his family. Jon is open to challenge and is a fast learner.

JON: Boy, this session has been an eye-opener for me. I've really been blind. My wife and kids don't see my investment—rather, my overinvestment—in work as something I'm doing for them. I've been fooling myself, telling myself that I'm working hard to get them the good things in life. In fact, I'm spending most of my time at work because I like it. My work is mainly for me. It's time for me to realign some of my priorities.

The last statement is a good intention, an indication on Jon's part that he wants to do something about a problem now that he sees it more clearly. It may be that Jon will now go out and put a different pattern of behavior in place without further help from the counselor. Or he may need some help in realigning his priorities.

- *Broad goals.* A broad goal is more than a good intention. It has content: it identifies the area in which the client wants to work and makes some general statement about that area. Let's return to the example of Jon and his overinvestment in work.

JON: I don't think I've gotten so taken up with work deliberately. That is, I don't think I'm running away from family life. But family life is deteriorating because I'm just not around enough. I must spend more time with my wife and kids. Actually, it's not just a case of must. I want to.

This is more than a declaration of intent because it says in a general way what Jon wants to do: "spend more time with my wife and kids." However, Jon's statement is a broad, rather than a specific, goal because it does not specify precisely what the new pattern of behavior will look like.

- *Specific goals.* Needs and wants can also be stated in terms of what a client wants to put in place in order to manage a problem situation or develop an opportunity. To move Jon toward greater specificity, the counselor asks him such questions as "What will your life with family look like when you make the change?"

JON: I'm going to consistently spend three out of four weekends a month at home. During the week I'll work no more than two evenings.

Notice how much more specific this statement is than "I'm going to spend more time with my family." Here Jon sets a goal in terms of a specific pattern of behavior he wants to put in place. He will reach his goal when he is habitually spending the indicated time with his family.

Helping clients move from good intentions to more and more specific goals is a shaping process. Consider the example of a couple whose marriage has degenerated into constant bickering, especially about finances.

- *Good intention*: We want to straighten out our marriage.
- *Broad goal*: We want to handle our decisions about finances in a much more constructive way.
- *Specific goal*: We try to solve our problems about family finances by fighting and arguing. We'd like to reduce the number of fights we have and increase the number of times we make decisions about money by talking them out. We're going to set up a month-by-month budget. We will have next month's budget ready the next time we meet with you.

These two people want to change their style of dealing with finances. More specifically, they want to replace one destructive pattern of behavior with a more constructive pattern. The goal is accomplished when the new pattern is in place. Declarations of intent, broad goals, and specific goals can all drive the kind of outcome-oriented and wise action that keeps the change process alive. Is it possible to get clients to be too specific about their goals? Yes, if it robs them of their spontaneity. Napoleon once said, "He will not go far who knows from the first where he is going."

Outcomes that can be verified. If the goal is clear enough, the client will be able to determine progress toward the goal. For many clients, being able to measure progress is an important incentive. If goals are stated too broadly, it is difficult to determine progress and accomplishment. "I want to have a better relationship with my wife" is a very broad goal, difficult to verify. "I want to spend more time with her" comes closer, but it is not clear what "more time" means.

It is not always necessary to count things in order to determine whether a goal has been reached, although sometimes counting is helpful. At a minimum, however, desired outcomes must be verifiable in some way. For instance, a couple might say something like "Our relationship is better, not because we've found ways of spending more time together, but because the quality of our time together has improved. We listen better, we talk about more personal concerns, we are more relaxed, and we make more mutual decisions about issues that affect us both, such as finances." The couple's accomplishment is one they can verify because they have spelled out what they mean by quality.

Help clients establish goals that make a difference, outcomes with impact.   The goals clients choose should have substance to them, the right kind of impact on their lives. First of all, to have substance, a goal must be related to managing the original problem situation. One client prided herself on being a "doer." She consistently set goals and acted to accomplish them. In the helping interviews it became increasingly clear that a rather acerbic interpersonal style was one of her main problems. But, since she insisted that her main problem was self-image, "an improved self-image" was one of her broad goals. She decided that looking more fashionable and getting a degree would help her with her self-image. She accomplished these goals, but she was still very dissatisfied with life. Perhaps she thought, at some level of her being, that a better self-image would take care of everything. Reflecting on this case some months after she terminated the helping process, her counselor mused, "I let her get away with murder."

Second, goals have substance to the degree that they help clients stretch themselves a bit. There is a good deal of research (Locke & Latham, 1984; Locke, Shaw, Saari, & Latham, 1981) suggesting that, other things being equal, harder goals can be more motivating than easier goals.

> Extensive research . . . has established that, within reasonable limits, the . . . more challenging the goal, the better the resulting performance. . . . People try harder to attain the hard goal. They exert more effort. . . . In short, people become motivated in proportion to the level of challenge with which they are faced. . . . Even goals that cannot be fully reached will lead to high effort levels, provided that partial success can be achieved and is rewarded. (Locke & Latham, 1984, pp. 21, 26)

I met an AIDS patient who was, in the beginning, full of self-loathing and despair. Eventually, however, he painted a new scenario in which he saw himself, not as a victim of his own lifestyle, but rather as a helper to other AIDS patients. Until close to the time of his death, he worked hard, within the limits of his physical disabilities, seeking out other AIDS sufferers, getting them to join self-help groups, and generally helping them to manage an impossible situation in a more humane way. When he was near death, he said that the last two years of his life, though at times they were very bitter, were among the best years of his life. He had set his goals high, but they proved to be quite realistic. At times clients who find change difficult creep up on goals that have more substance to them. At first they settle for an inadequate goal and then move on at their own pace to a more substantial goal.

Help clients formulate realistic goals.   Setting stretch goals can help clients energize themselves. They rise to the challenge. On the other hand, goals set too high do more harm than good. Locke and Latham (1984) put it succinctly: "Nothing breeds success like success. Conversely, nothing causes feelings of despair like perpetual failure. A primary purpose of goal setting is to increase the motivation level of the individual. But goal setting can have precisely the opposite effect if it produces a yardstick that constantly makes the individual feel inadequate" (p. 39).

Wheeler and Janis (1980) caution against the search for the "absolute best" goal all the time. It is not realistic: "Sometimes it is more reasonable to choose a satisfactory alternative than to continue searching for the absolute best. The time, energy, and expense of finding the best possible choice may outweigh the improvement in the choice" (p. 98). Consider the following case.

> Joyce, a flight attendant nearing middle age, centered most of her non-flying life around her aging mother. Her mother had been pampered by her now-deceased husband and her three children and allowed to have her way all her life. Her mother now played the role of the tyrannical old woman who constantly feels neglected and who can never be satisfied. Though Joyce knew that she could live much more independently without abandoning her mother, she found it very difficult to make choices for herself. Guilt stood in the way of any change in her relationship with her mother. She even said that being a virtual slave to her mother's whims was not as bad as the guilt she experienced when she stood up to her mother or "neglected" her.
>
> The counselor helped Joyce experiment with a few new ways of dealing with her mother. For instance, Joyce went on a two-week trip with friends even though her mother objected, saying that it was ill-timed. Although the experiments were successful in that no harm was done to Joyce's mother and Joyce did not experience excessive guilt, counseling did not help her restructure her relationship with her mother in any substantial way. The experiments did give her a sense of greater freedom. For instance, she felt freer to say no to this or that demand on the part of her mother. This provided enough slack, it seems, to make Joyce's life more livable.

In this case, counseling helped the client fashion a life that was "a little bit better," though not as good as the counselor thought it could be. When asked, "What do you want?" Joyce had in effect replied, "I want more freedom but I do not want to abandon my mother." Joyce's "new" scenario did not differ dramatically from the old. But perhaps it was enough for her. It was a case of choosing a satisfactory alternative rather than the best.

Needs and wants translated into goals must be realistic. A goal is realistic if the client has access to the resources needed to accomplish it, external circumstances do not prevent its accomplishment, the goal is under the client's control, the goal is sustainable, and the benefits outweigh the costs.

Resources: Help clients choose goals for which the resources are available. It does little good to help clients develop specific and substantive and verifiable goals if the resources needed for their accomplishment are not available. Consider the case of John, a college graduate who took six years to get his degree because he "messed around a bit" and because of limited intellectual resources.

INSUFFICIENT RESOURCES: John wants to become a fully-licensed clinical psychologist. But he lacks the intellectual talent and the financial resources needed to get into graduate school.

SUFFICIENT RESOURCES: John changes his focus to becoming a rehabilitation counselor. He gets a job and, while working, takes an evening graduate course in rehabilitation counseling to see whether he is really interested in this field. He also spends one night a week at a rehabilitation clinic to determine whether he is capable of doing the work and whether he likes it.

Obstacles: Help clients choose goals not blocked by environmental obstacles. A goal is not really a goal if there are environmental obstacles that prevent its accomplishment.

Jessie felt like a second-class citizen at work. He thought that his supervisor gave him most of the dirty work and that there was an undercurrent of prejudice against Hispanics in the plant. He wanted to quit and get another job, one that would pay the same relatively good wages. However, the country was deep into a recession, and there were practically no jobs available in the area where he worked. For the time being, then, his goal was not workable. He needed an interim goal related to coping with his present situation. Since he wanted to become more skilled to prepare himself for another job, he volunteered for special assignments and worked extra hours for no pay in order to learn new skills. He felt good about what he was learning and more easily ignored the prejudice.

Of course, not every obstacle is insurmountable. Skilled counselors help clients distinguish between obstacles that need to be overcome and goal-blocking obstacles.

Control: Help clients choose goals that are under their control. Sometimes clients defeat their own purposes by setting goals that are not under their control. For instance, it is common for people to believe that their problems would be solved if only other people would not act the way they do. In most cases, however, we do not have any direct control over the ways others act. Consider the following example.

Tony, a 16-year-old boy, felt that he was the victim of his parents' inability to relate to each other. Each tried to use him in the struggle, and at times he felt like a ping-pong ball. A counselor helped him see that he could probably do little to control his parents' behavior, but that he might be able to do quite a bit to control his reactions to his parents' attempts to use him. For instance, when his parents started to fight, he could simply leave instead of trying to "help." If either tried to enlist him as an ally, he could say that he had no way of knowing who was right. Tony also worked at creating a good social life outside the home. This helped him weather the tension he experienced there.

Tony found a new way of managing his interactions with his parents in order to minimize their attempts to use him as a pawn in their own interpersonal game.

Sustainability and flexibility: Help clients set goals that can be sustained. Clients need to commit themselves to goals with staying power. One separated couple said that they wanted to get back together again. They did so only to separate and get divorced within six months. Their goal of getting back together again was achievable but not sustainable. Perhaps they should have said something like this: "What do we need to do, not only to get back together, but to stay together? What would our marriage have to look like?"

Goals are more sustainable if they are also flexible. One way of helping clients increase flexibility is to help them choose broad goals that have specific goal alternatives.

> Bob, a gay man, had been somewhat promiscuous before the arrival of the AIDS epidemic. He said to a counselor, "I'm so grateful that I dodged the bullet myself. Now I want to do something to help." The counselor helped him translate this statement of intent into a broad goal of "doing volunteer work that would make a difference in people's lives." Bob's initial instinct was to work directly with AIDS patients, but the counselor knew how grueling and depressing that could be. Bob, with some guidance, finally chose to work for a clinic that provided a broad range of services—from working directly with dying patients to fund raising to education and prevention programs. The wisdom of the choice soon became clear to Bob. He did find working directly with patients, especially those with advanced cases of the disease, very difficult. He did "his share"—and he found that he could contribute to the cause by helping out with the clinic's other activities. He "made a difference" in the lives of patients—and in the lives of teenagers through prevention programs.

Counseling is a living, organic process. Just as organisms adapt to their changing environments, the choices clients make need to be adapted to their changing circumstances. If Bob had taken an all-or-nothing-approach by committing himself just to working directly with patients, he might have abandoned his efforts.

Cost-Benefit ratio: Help clients set goals that don't cost more than they are worth. Some goals that can be accomplished carry too high a cost in relation to the payoff. It may sound overly technical to ask whether any given goal is cost effective, but the principle remains important. Skilled counselors help clients budget, rather than squander, their resources.

> Eunice discovered that she had a terminal illness. In talking with several doctors, she found out that she would be able to prolong her life a bit through a combination of surgery, radiation treatment, and chemotherapy. However, no one suggested that these would lead to a cure. She also found out what each of these three forms of treatment and each combination would cost, not so much in terms of money, but in terms of added anxiety and pain. Ultimately she decided against all three, since no combination of them promised much in terms of the quality of the life that was being prolonged. Instead, with the help of a doctor who was an expert in hospice care, she developed a scenario that would ease her anxiety and her physical pain as much as possible.

Another patient might have made a different decision. Costs and payoffs are relative. Some clients might value an extra month of life no matter what the cost.

Since it is often impossible to determine the cost-benefit ratio of any particular goal, counselors can add value by helping clients understand the *consequences* of choosing any given goal. For instance, if a client sets her sights on a routine job with minimally adequate pay, this outcome might well take care of some of her immediate needs but prove to be a poor choice in the long run. Helping clients foresee the consequences of their choices may not be easy. Another woman with cancer felt she was no longer able to cope with the sickness and depression that came with her chemotherapy treatments. She decided abruptly one day to end the treatment, saying that she didn't care what happened. No one helped her explore the consequences of her decision. Eventually, when her health deteriorated, she had second thoughts about the treatments, saying, "There are still a number of things I must do before I die." But it was too late. Some challenge on the part of a helper might have helped her make a better decision.

Help clients choose goals consistent with their values.   Although helping is a process of social influence, it remains ethical only if it respects, within reason, the values of the client. Values are criteria we use to make decisions. While helpers may challenge clients to reexamine their values, they should in no way encourage clients to perform actions that are not in keeping with their values. For instance, the son of Antonio and Consuela Garza is in a coma in the hospital after an automobile accident. He needs a life-support system to remain alive. His parents are experiencing a great deal of uncertainty, pain, and anxiety. They have been told that there is practically no chance that their son will ever come out of the coma. The counselor should not urge them to terminate the life-support system if that action is counter to their values. However, the counselor can help them explore and clarify the values involved. In this case the counselor suggests that they discuss their decision with their clergyman. In doing so, they find out that the termination of the life-support system would not be against the tenets of their religion. Now they are free to explore other values that relate to their decision.

Some problems involve clients trying to pursue contradictory goals or values. Dillard, the ex-Marine, wanted to get an education, but he also wanted to make a decent living as soon as possible. The former goal would put him in debt, but failing to get a college education would lessen his chances of securing the kind of job he wanted. In this case the counselor helped him identify and use his values to consider some tradeoffs. Dillard chose to work part-time and go to school part-time. He chose a job in an office instead of a job in construction. Even though the latter paid better, it would be much more exhausting and would leave him with little energy for school.

Help clients establish realistic time frames for the accomplishment of goals.   Goals that are to be accomplished "sometime" probably won't be accomplished at all. Therefore, helping clients put some time frames in their goals can add value. Greenberg (1986) talked about immediate outcomes, intermediate change, and final outcomes.

- *Immediate* outcomes are changes evident in the helping sessions themselves.
- *Intermediate* outcomes are changes in attitudes and behaviors that lead to further change.
- *Final* outcomes refer to the completion of the overall program for constructive change through which problems are managed and opportunities developed.

Jensen, a 22-year-old on probation for shoplifting, was seeing a counselor as part of a court-mandated program. An immediate need in his case was overcoming his resistance to his court-appointed counselor and developing a working alliance. Because of the counselor's skill and her unapologetic caring attitude that had some toughness in it, he quickly came to see her as "on his side." An intermediate outcome was attitudinal. Brainwashed by what he saw on television, he thought that America owed him some of its affluence and that personal effort had little to do with it. The counselor helped him see that his entitlement attitude was unrealistic and that hard work played a key role in most payoffs. There were two significant final outcomes in Jensen's case. First, he made it through the probation period free of any further shoplifting attempts. Second, he acquired and kept a job that helped him pay his debt to the retailer.

Taussig (1987) talked about the usefulness of setting and executing mini-goals early in the helping process. Consider the case of Terry.

> Terry, a 16-year-old school dropout and loner, was arrested for arson. Though he lived in the inner city and came from a single-parent household, it was difficult to discover just why he had turned to arson. He had torched a few structures that seemed relatively safe to burn. No one was injured. Was his behavior a cry for help? Was it social rage expressed in vandalism? Was it just a way of getting some kicks? The social worker assigned to the case found these questions too speculative to be of much help. Instead of looking for the root causes of Terry's malaise, she tried to help him set some simple goals that appealed to him and that could be accomplished relatively quickly. One goal was social support. The counselor helped Terry join a social club at a local church. A second goal was having a role model. Terry struck up a friendship with one of the more active members of the club, a dropout who had gotten a high-school equivalency degree. He also received some special attention from one of the adult monitors of the club. This was the first time he had experienced the presence of a strong adult male in his life. A third goal was broadening his view of the world. A group of seminarians who worked in both the black and white communities invited Terry and a couple of the other boys to help them in a white parish. This was the first time he had been out of the ghetto and helped him push back the walls a bit. The accomplishment of these mini-goals fulfilled some of Terry's needs and helped him experience himself, others, and the world about him in much more constructive ways.

It is not suggested here that goal setting is a facile answer to intractable social problems. But the achievement of sequenced mini-goals can go a long way toward making a dent in intractable problems.

There is no such thing as a set time frame for every client. Some goals need to be accomplished now, some soon; others are shorter-term goals; still others are long-term. Consider Dillard, our ex-Marine.

- A *"now" goal*: Some immediate relief from debilitating anxiety attacks, followed by a gradual decrease in the number and severity of the attacks.
- A *"soon" goal*: A pattern of better care of his personal appearance consistently in place.
- A *shorter-term goal*: Decisions made with respect to the mix of school and work.
- A *longer-term goal*: The fashioning of a career and prospects for marriage and a family.

There is no particular formula for helping all clients choose the right mix of goals at the right time and in the right sequence. Helping is based on problem-management principles, yet it remains an art.

In general, to be workable, goals must meet all six requirements described in this chapter: they must be truly goals, outcomes, and accomplishments, and not just sets of activities; they must be specific enough to drive behavior; they must be challenging enough to make a difference; they must be doable and sustainable; they should challenge but should not violate the client's values; and they must be set in a reasonable time frame. Failure to meet even one of the six criteria may prove to be the fatal flaw in a client's movement toward problem-managing action.

It is not always necessary, then, to make sure that each goal in a client's program for constructive change has all six characteristics. For some clients, identifying broad goals is enough. For others, some help in formulating more specific goals is called for. The principle is clear: help clients develop goals that have some probability of success. In one case, this may mean helping a client deal with clarity; in another, with substance; in still another, with realism, values, or time frame. The balance-sheet methodology can help clients choose goals that make a difference.

STEP II-C: WHAT ARE CLIENTS WILLING TO PAY FOR WHAT THEY WANT?

As mentioned at the beginning of this chapter, Step II-C is not really a step in the true sense of the term but a dimension of the goal-setting process. Clients can formulate goals without being willing to pay for them. Once clients state what they want and set goals, the drama is set, as it were. It is as if the client's old self or old lifestyle begins vying for resources with the client's new self or new lifestyle.

Commitment

History is full of examples of how the will to accomplish some goal is so strong that people have been able to accomplish seemingly impossible things.

> A woman with two sons in their twenties was dying of cancer. The doctors thought she could go at any time. However, one day she told the doctor that she wanted to live to see her older son get married in six months' time. The doctor talked vaguely about "trusting in God" and "playing the cards she had been dealt." Against all odds the woman lived to see her son get married. Her doctor was at the wedding. During the reception he went up to her and said, "Well, you got what you wanted. Despite the way things are going, you must be deeply satisfied." She looked at him wryly and said, "But, Doctor, my second son will get married some day."

While the job of counselors is not to encourage clients to heroic efforts, counselors should not undersell clients either.

In this step, which is usually intermingled with the other two steps of Stage II, counselors help their clients pose and answer such questions as:

- Why should I pursue this goal?
- Is it worth it?
- Is this where I want to invest my limited resources of time, energy, and money?
- What competes for my attention?
- What are the incentives for pursuing competing agendas that would sap energy from my change program?
- How strong are these competing agendas?

Figure 13-2 adds this final step to Stage II. Again, there is no formula. Some clients, once they establish goals, race to accomplish them. At the other end of the spectrum are clients who, once they decide on goals, stop dead in the water. Further, the same client might speed toward the accomplishment of one goal and drag her feet on another. The job of the counselor is to help clients face up to their commitments.

Helping Clients Commit Themselves

There is a difference between initial commitment to a goal and an ongoing commitment to a strategy or plan to accomplish the goal. The proof of initial commitment is some movement to action. For instance, one client who

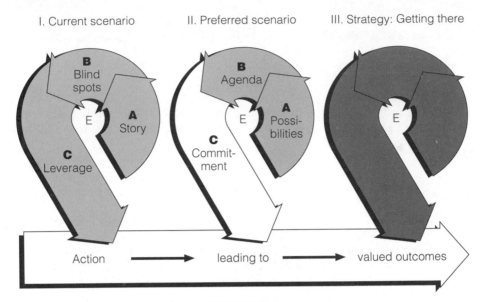

FIGURE 13-2
Step II-C: Helping Clients Commit to Agendas

chose as a goal a less abrasive interpersonal style began to engage in an "examination of conscience" each evening to review what his interactions with people had been like that day. In doing so, he discovered, somewhat painfully, that in some of his interactions he came near to contempt of the other. "Wait now," he said to himself, "What's going on here? Have I let myself go this far?" Ongoing commitment, on the other hand, takes the form not only of steady progress toward goals but also of getting up and moving forward after failures. This chapter deals with initial commitment. Chapter 16 deals with ongoing commitment.

There are a range of things you can do to help clients in their initial commitment to goals and the kind of action that is a sign of that commitment. Counselors can help clients by helping them make goals appealing, by helping them enhance their sense of ownership, and by helping them deal with competing agendas.

Appeal: Help clients set appealing goals.   Although setting appealing goals seems to be common sense, it is not always easy to do. For instance, for many—if not most—addicts, a drug-free life is far from being immediately appealing.

> Ed tries to help Dillard work through his resistance to giving up prescription drugs. He listens and is empathic; he also challenges the way Dillard has come to think about drugs and his dependency on them. One day Ed says something about "giving up the crutch and walking straight." In a flash Dillard sees

himself not as addicted to drugs but as a cripple. Some of the soldiers of Desert Storm had become cripples. He knew that they longed for the days when they could throw their crutches away. The image of "throwing away the crutch" and "walking straight" proved to be very appealing to Dillard.

A goal is appealing if there are incentives for pursuing it. Counselors need to help clients in their search for incentives. Ordinarily, negative goals — giving something up — need to be translated into positive goals — that is, into getting something. It was much easier for Dillard to commit himself to returning to school than to giving up prescription drugs, because school represented something he was getting. Images of himself with a degree and of holding some kind of professional job were solid incentives. The picture of him throwing away the crutch proved to be an important incentive in cutting down on drug use.

Ownership: Make sure that the goals clients set are their own.   Not only should goals be in keeping with clients' values, but the clients should own them. It is essential that the goals chosen be the clients' goals rather than the helper's or someone else's. Various kinds of probes can be used to help clients discover what they want to do in order to manage some dimension of a problem situation more effectively. For instance, in a film of a counseling session, Rogers is asked by a woman what she should do about her relationship with her daughter (Rogers, Perls, & Ellis, 1965). He says to her, "I think you've been telling me all along what you want to do." She knew what she wanted the relationship to look like, but she was asking for his approval. If he had given it, the goal would, to some degree, have become his goal instead of hers. At another time he asks: "What is it that you want me to tell you to do?" This question puts the responsibility for goal setting where it belongs — on the shoulders of the client.

> Cynthia was dealing with a lawyer because of an impending divorce. They had talked about what was to be done with the children, but no decision had been reached. One day she came in and said that she had decided on mutual custody. She wanted to work out such details as which residence, hers or her husband's, would be the children's principal one and so forth. The lawyer asked her how she had reached a decision. She said that she had been talking to her husband's parents, with whom she was still on good terms, and that they had suggested this arrangement. The lawyer challenged Cynthia to take a closer look at her decision. "Let's start from zero," he said, "and you tell me what arrangements you want." He did not want to help her carry out a decision that was not her own.

Choosing goals suggested by others enables clients to blame others if they fail to reach the goals.

Initial commitment can take different forms. The least useful is *mere compliance*. "Well, I guess I'll have to change some of my habits if I want to keep my marriage afloat" does not augur well for ongoing commitment. But it may be better than nothing. *Buy in* is a level up from compliance. "Yes, restricting my sex life does make sense. After all, our marriage and family life are so much bigger than sex." This client has moved beyond mere compliance. But sometimes mere buy in does not provide enough staying power because it depends too much on reason. "This is logical" is far different from "This is what I really want!" *Ownership* is a higher form of commitment. It means that the client can say, "This goal is not someone else's, it's not just a good idea; it is mine, it is what I want to do." Consider the following case.

> A counselor worked with a manager whose superiors had intimated that he would not be moving much further in his career unless he changed his style in dealing with the workers in his unit and other key people with whom he had interactions within the organization. At first the manager resisted setting any goals. His initial response was: "What they want is a lot of hogwash. It won't do anything to make the business better."
>
> One day, when asked whether accomplishing what "they" wanted would cost him that much, he pondered a few moments and then said, "No, not really." This got him started. With a bit of help from the counselor he identified a few areas of his managerial style that could well be "polished up." He had moved beyond resistance to the compliance level. Within a year he got much more into the swing of things. Given the favorable response he had gotten from the people who reported to him, he was able to say, "Well, I now see that this makes sense. But I'm doing it because it has a positive effect on the people in the department. It's the right thing to do." Buy in had arrived. A year later he moved up another notch. He became much more proactive in finding ways to improve his style. He delegated more, gave people feedback, asked for feedback, held a couple of managerial retreats, joined a human resource task force, and routinely rewarded his direct reports for their successes. Now he began to say such things as, "This is actually fun." Ownership had arrived. The people in his department began to see him as one of the best executives in the company. This process took over two years.

The manager did not change his opinion of some of his superiors. He was right in pointing out that they didn't observe their own rules.

The use of contracts to structure the helping process itself was discussed in Chapter 3. Self-contracts, contracts that clients make with themselves, can also help clients commit themselves to new courses of action. Contracts are promises clients make to themselves to behave in certain ways and to attain certain goals; they are also ways of making goals more focused. It is not only the expressed or implied promise that helps, but the explicitness of the commitment.

> About a month after one of Dora's two young sons disappeared, she began to grow listless and depressed. She was separated from her husband at the time the boy disappeared. By the time she saw a counselor a few months later, a pattern of depressed behavior was quite pronounced. Her conversations with the counselor helped ease her feelings of guilt—she stopped engaging in self-

blaming rituals — yet she remained listless. She shunned relatives and friends, kept to herself at work, and even distanced herself emotionally from her other son. She resisted developing images of a better future, because the only better future she would allow herself to imagine was one in which her son had returned.

Some strong challenging from Dora's sister-in-law, who visited her from time to time, helped jar her loose from her preoccupation with her own misery. "You're trying to solve one hurt, the loss of Bobby, by hurting Timmy and hurting yourself. I can't imagine in a thousand years that that's what Bobby would want!" her sister-in-law screamed at her one night. Afterwards Dora and the counselor discussed a "recommitment" to Timmy, to herself, and to their home. Through a series of contracts she began to reintroduce patterns of behavior that had been characteristic of her before the tragedy. For instance, she contracted to opening her life up to relatives and friends once more, to creating a much more positive atmosphere at home, to encouraging Timmy to have his friends over, and so forth. Contracts worked for Dora because, as she said to the counselor, "I'm a person of my word."

When Dora first began implementing these goals, she felt she was just going through the motions. However, what she was really doing was acting herself into a new mode of thinking. Contracts helped Dora in both her initial commitment to a goal and in her movement to action. In counseling, contracts are not legal documents but human instruments to be used if they are helpful. They often provide the structure and the incentives some clients need.

Priorities: Helping clients deal with competing agendas.   Clients often set goals and formulate programs for constructive change without taking into account competing agendas. For instance, one manager wanted to change his overly-critical interpersonal style, but crises in the economy and a divorce set up competing agendas and sapped his resources. None of the goals of his interpersonal-style agenda was accomplished. Programs for constructive change involve a rearrangement of priorities. If a client is to be a full partner in the re-invention of his marriage, then he cannot spend as much time with the boys. The unemployed white collar worker might have to put aside habitual interests and devote himself full time to a search for work. One executive, out of work for over nine months, was pulled up short in a self-help group when one of his peers said, "You're still living your old life and just playing at getting a job."

This is not to say that competing agendas are frivolous. The woman who wants to expand her horizons by getting involved in social settings outside the home has to figure out how to handle the tasks at home. They are strong competitors. The single parent who wants a promotion at work needs to balance her new responsibilities with involvement with her children. There is just so much time, energy, and other resources. A counselor who had worked with a two-career couple as they made a decision to have a child helped them think of competing agendas once the pregnancy started. A year after the baby was born they saw the counselor again for a couple of sessions to work on some issues that had come up. They started the session by saying, "We are glad that you talked about competing agendas when we were struggling with the decision to become parents. In our discussions with one another we have gone back time and time again and reviewed what we said and what we decided. It helped stabilize us for the last two years."

THE SHADOW SIDE OF CHOICE AND COMMITMENT

The shadow side of Stage II is found mainly in Steps II-B and II-C — choice and commitment — because they involve decision making. The shadow side of decision making, discussed in Chapter 10, makes itself felt here. Leaving the safe harbor of problem and opportunity exploration in order to move to commitment is not easy for many clients. Nor is it easy for some counselors who feel competent and comfortable in the exploration stage but somewhat at sea in trying to help clients move on to programs for constructive change. Helpers who feel competent in Stages II and III stray in other ways. Without being aware of it, they begin to push their own values even though they have remained neutral during the exploration stage. Other helpers become a bit uncomfortable at some level of their being as they see their clients "growing up" and becoming more independent.

A cynic once said, "The only thing worse than not getting what you want is getting what you want." The responsibilities accompanying getting what you want — a drug-free life, a renewed marriage, custody of the children, a promotion, the peace and quiet of retirement, freedom from an abusive husband — often cause a new set of problems. It is one thing for parents to decide to give their children more freedom and then freeze as they watch them use this freedom. The fact that choosing what one wants and translating these choices into goals can lead to a new set of problems is vaguely understood by clients and may keep them from making choices and commitments. Furthermore, there is a phenomenon called post-decisional depression. Once choices are made, clients begin to have second thoughts, which often keep them from acting on their decisions.

Even self-contracts have a shadow side. There is no such thing as a perfect contract. Most people don't think through the consequences of all the provisions of a contract, whether it be marriage, employment, or self-contracts designed to enhance a client's commitment to goals. And even people of goodwill unknowingly add covert codicils to contracts they make with themselves and others; for example, "I'll pursue this goal until it begins to cost me money" or "I won't be abusive unless she pushes me to the wall." The codicils are buried deep in the decision-making process and only gradually make their way to the surface.

STAGE II AND ACTION

The work of Step II-A is just what some clients need. It frees them from thinking solely about problem situations and unused resources and enables them to begin fashioning better futures. Once they identify some of their wants and needs, they move into action. Other clients move into action too quickly. The focus on the future liberates them from the past and the first few possibilities are very attractive. They need the kind of focus and direction provided by Step II-B. Weighing alternatives and shaping goals help them move to action that is both directed and prudent. For still other clients, the search for incentives for commitment is the trigger for action. Once they see what's in it for them—a kind of upbeat and productive selfishness, if you will—they are ready to work.

In practice, however, the three steps of Stage II overlap, just as Stage II itself overlaps with the other two stages. This is illustrated in Figure 13-3.

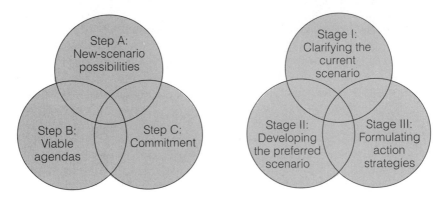

FIGURE 13-3
Overlays in the Steps and Stages of the Helping Process

- To what degree am I helping clients choose specific goals from among a number of preferred-scenario possibilities?
- To what extent are the goals set by each client
  - stated as outcomes and accomplishments rather than activities?
  - clear, specific, and verifiable?
  - realistic in terms of resources, obstacles, control, sustainability, and benefits?
  - substantive in terms of managing the original problem situation and helping the client stretch?
  - substantive in terms of having a positive impact on the client's life?
  - in keeping with the client's values?
  - set in a reasonable time frame?
- How well do I challenge clients to translate good intentions into broad goals and broad goals into specific, actionable goals?
- How effectively do I help clients explore the consequences of the goals they are setting?

- What do I do to make sure that the goals clients are setting are really their goals and not mine or those of a third party?
- In what ways do I help clients focus on the appealing dimensions of the goals being set?
- How effectively do I perceive and deal with the misgivings clients have about the goals they are formulating?
- To what degree do I help clients enter into self-contracts with respect to the accomplishment of goals?
- What do I do to help clients move to initial goal-accomplishing action?

GETTING THERE—HELPING CLIENTS IMPLEMENT THEIR GOALS

Setting substantial goals (Stage II) is not a problem if realistic strategies for implementing them are identified and carried out (Stage III). The two together constitute the client's program for constructive change. Stage II deals with *what* clients want; Stage III deals with *how* to get there. Generally speaking, the clearer preferred-scenario goals are, the easier it is to find strategies for accomplishing them.

Step III-A. Help clients develop strategies for accomplishing their goals.

Step III-B. Help clients choose strategies tailored to their preferences and resources.

Step III-C. Help clients formulate actionable plans.

Stage III adds the final pieces to a client's program for constructive change; it deals with the game plan. However, these three steps constitute planning for action and should not be confused with action itself. Without action, a program for constructive change is nothing more than a wish list. Unfortunately, too many programs are just that.

As in Stage II, the three steps of Stage III can be translated into the ordinary language of the client.

Step III-A. Action Strategies—What do I need to do to get what I want?

Step III-B. Choices—What actions are best for me?

Step III-C. Plans—What do I need to do first? Second?

Finally, as in Stage II, these steps, although rationally distinct, are intermingled in practice.

CHAPTER 14

STEP III-A: STRATEGIES — WHAT DO CLIENTS NEED TO DO TO GET WHAT THEY WANT?

HELPING CLIENTS DEVELOP STRATEGIES TO GET WHAT THEY WANT AND NEED

Avoiding Imprudent Action

Delaying Action

HOW MANY WAYS ARE THERE FOR CLIENTS TO GET WHAT THEY WANT AND NEED?

Help Clients Brainstorm Strategies for Accomplishing Goals

Develop Frameworks for Stimulating Clients' Thinking about Strategies

WHAT SKILLS DO CLIENTS NEED TO GET WHAT THEY WANT?

LINKING STRATEGY FORMULATION TO ACTION

EVALUATION QUESTIONS FOR STEP III-A

I. Current scenario   II. Preferred scenario   III. Strategy: Getting there

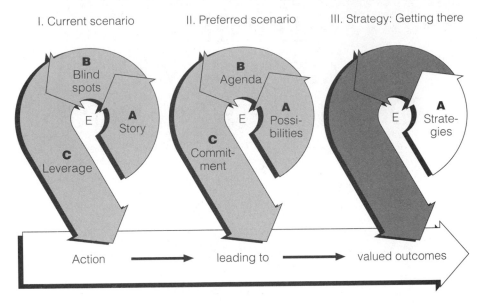

FIGURE 14-1
Step III-A: Developing Action Strategies

Developing a range of possible strategies to accomplish goals, Step III-A of the helping process illustrated in Figure 14-1, is a powerful exercise. Clients who feel hemmed in by their problems and unsure of the viability of their goals are liberated through this process.

HELPING CLIENTS DEVELOP STRATEGIES TO GET WHAT THEY WANT AND NEED

Strategy is the art of identifying and choosing realistic courses of action for achieving goals—and doing so under adverse conditions, such as war. The problem situations in which clients are immersed constitute adverse conditions; clients often are at war with themselves and the world around them. Helping clients develop strategies to achieve goals can be a most thoughtful, humane, and fruitful way of being with them. This step in the counseling process is another that helpers sometimes avoid because they consider it too technological. They do their clients a disservice. Clients with goals but no clear idea of how to accomplish them still need help. Step III-A is important both for clients who rush into action and for clients who don't know what to do.

Avoiding Imprudent Action

For many clients, the problem is not that they refuse to act. Rather, they act without direction or they act imprudently. Rushing off to try the first strategy that comes to mind is often imprudent.

Elmer injured his back and underwent a couple of operations. After the second operation he felt a little better, but then his back began troubling him again. When the doctor told him that further operations would not help, Elmer was faced with the problem of handling chronic pain. It soon became clear that his psychological state affected the level of pain. When he was anxious or depressed, the pain always seemed much worse.

Elmer was talking to a counselor about all of this when he read about a pain clinic located in a western state. Without consulting anyone, he signed up for a six-week program. Within ten days he was back, feeling more depressed than ever. He had gone to the program with extremely high expectations because his needs were so great. The program was a holistic one that helped the participants develop a more realistic lifestyle. It included programs dealing with nutrition and quality of interpersonal life. Group counseling was part of the program, and the groups included a training approach. For instance, the participants were trained in behavioral approaches to the management of pain. The trouble was that Elmer had arrived at the clinic, which was located on a converted farm, with unrealistic expectations. He had not really studied the materials that the clinic had sent to him. Since he had expected to find marvels of modern medicine that would magically help him, he was extremely disappointed when he found that the program focused mainly on reducing and managing rather than eliminating pain.

Elmer's goal was to be completely free of pain, and he failed to see that this might not be possible. A more realistic goal would have centered around the reduction and management of pain. Elmer's helper failed to help him avoid two mistakes: not setting a realistic goal and, in desperation, seizing the first strategy that came along.

Delaying Action

At the other end of the spectrum are clients who keep putting action, at least goal-directed action, off. While there are many reasons for procrastination, not knowing what to do is one of the more important. Some clients, once they have a clear idea of what they want to do to manage some problem situation, mobilize whatever resources are necessary to achieve their goals. Other clients, however, even though they have a fairly clear understanding of the problem situation and know where they want to go, still lack a clear idea of how to get there.

Eula, disappointed with her relationship with her father in the family business, decided that she wanted to go out on her own. She thought that she could capitalize not only on the business skills she had picked up in school and in the family business but also on what she had learned, sometimes painfully, about the ways in which family dynamics can pervade the business. Her goal, then, was to establish a consulting practice specializing in services to small family businesses. She sold some of the stock she had in the family business back to her father and became independent.

But, even though Eula was very busy, a year went by and she still did not have any clients. A counselor helped her see two things. First, her activities—researching the field, learning more about the psychology of family dynamics, going to seminars, getting involved for short periods with professionals such as accountants and lawyers who did a great deal of business with family-owned firms, even drawing up tentative business plans and drafts of a brochure — were all good in themselves but did little to create a business. There were a lot of activities, but her goal seemed as distant as ever. Second, the counselor helped Eula see that at some level of her being she was afraid of starting a new business. Over-preparation was her way of managing that fear. He helped her discuss and deal with her fears and he helped her identify strategies for getting her business moving.

Flaws in goal setting, lack of commitment to goals, and a reluctance to act all make it difficult for clients to develop strategies.

How Many Ways Are There for Clients to Get What They Want and Need?

Helping clients stimulate their imaginations and engage in divergent thinking is an important part of Step III-A. Most clients do not automatically seek different routes to goals and then choose from among them.

Help Clients Brainstorm Strategies for Accomplishing Goals

Brainstorming, discussed in Chapter 11, plays an important part in strategy development. The quality and efficacy of an action plan tend to be better if the program is chosen from among a number of possibilities. Consider the case of Karen, who had come to realize that heavy drinking was ruining her life. Her goal was to stop drinking. She felt that it simply would not be enough to cut down; she had to stop. To her the way forward seemed simple enough: whereas before she drank, now she wouldn't. Because of the novelty of not drinking, she was successful for a few days; then she fell off the wagon. This happened a number of times until she finally realized that she could use some help. Stopping drinking, at least for her, was not as simple as it seemed.

A counselor at a city alcohol and drug treatment center helped her explore a number of techniques that could be used in an alcohol-management program. Together they came up with the following possibilities:

- Just stop cold turkey and get on with life.
- Join Alcoholics Anonymous.
- Move someplace declared "dry" by local government.
- Take a drug that causes nausea if followed by alcohol.
- Replace drinking with other rewarding behaviors.
- Join some self-help group other than Alcoholics Anonymous.

- Get rid of all liquor in the house.
- Take the "pledge" not to drink; to make it more binding, take it in front of a minister.
- Join a residential hospital detoxification program.
- Avoid friends who drink heavily.
- Change other social patterns; for instance, find places other than bars and cocktail lounges to socialize.
- Try hypnosis to reduce the drive to drink.
- Use behavior modification techniques to develop an aversion for alcohol; for instance, pair painful but safe electric shocks with drinking or even thoughts about drinking.
- Change self-defeating patterns of self-talk, such as "I have to have a drink" or "One drink won't hurt me."
- Become a volunteer to help others stop drinking.
- Read books and view films on the dangers of alcohol.
- Stay in counseling as a way of getting support and challenge for stopping.
- Share intentions of stopping drinking with family and close friends.
- Spend a week with an acquaintance who does a great deal of work in the city with alcoholics, and go on his rounds with him.
- Walk around skid row meditatively.
- Have a discussion with members of the family about the impact drinking has on them.
- Discover things to eat that might help reduce the craving for alcohol.
- Get a hobby or avocation that captures the imagination.
- Substitute a range of self-enhancing activities for drinking.

This list contained many more items than Karen would have thought of had she not been stimulated by the counselor to take a census of possible strategies. One of the reasons that clients are clients is that they are not very creative in looking for ways of getting what they want. Once goals are established, getting them accomplished is not just a matter of hard work. It is also a matter of imagination.

If a client is having a difficult time coming up with strategies, the helper can prime the pump by offering a few suggestions. Driscoll (1984) put it well.

> Alternatives are best sought cooperatively, by inviting our clients to puzzle through with us what is or is not a more practical way to do things. But we must be willing to introduce the more practical alternatives ourselves, for clients are often unable to do so on their own. Clients who could see for themselves the more effective alternatives would be well on their way to using them. That clients do not act more expediently already is in itself a good indication that they do not know how to do so. (p. 167)

The helper may need to suggest alternatives, but can do so in such a way that the principal responsibility for evaluating and choosing possible strategies stays with the client. For instance, there is the "prompt and fade" technique. The counselor can say, "Here are some possibilities. . . . Let's review them and see whether any of them make sense to you or suggest further possibilities." Or, "Here are some of the things that people with this kind of problem situation have tried. . . . How do they sound to you?" The "fade" part of this technique keeps it from being advice giving: prompts are stated in such a way that the client has to work with the suggestion.

All the communication skills reviewed earlier in this book — especially prompts, probes, and invitations to self-challenge — are used to help clients develop strategies for action. Without taking over responsibility for the census of strategies, you can use certain probes and prompts to stimulate your clients' imaginations.

Develop Frameworks for Stimulating Clients' Thinking about Strategies

How can helpers find the right probes to help clients develop a range of strategies? Simple frameworks can help. Consider the following case.

> Jason has terminal cancer. He has been in and out of the hospital several times over the past few months, and he knows that he probably will not live more than a year. He would like the year to be as full as possible, and yet he wants to be realistic. He hates being in the hospital, especially a large hospital, where it is so easy to be anonymous. One of his goals is to die outside the hospital. He would like to die as benignly as possible and retain possession of his faculties as long as possible. How is he to achieve these goals?

Probes and prompts can be used to discover possible strategies by investigating possible resources in the client's life, including people, models, communities, places, things, organizations, programs, and personal resources.

Individuals.   What individuals might help clients achieve their goals? Jason gets the name of a local doctor who specializes in the treatment of chronic cancer-related pain. The doctor teaches people how to use a variety of techniques to manage pain. Jason says that perhaps his wife and daughter can learn how to give simple injections to help him control the pain. Also, he thinks that talking every once in a while with a friend whose wife died of cancer, a man he respects and trusts, will help him find the courage he needs.

Models and exemplars.   Does the client know people who are presently doing what he or she wants to do? One of Jason's fellow workers died of cancer at home. Jason visited him there a couple of times. That's what gave him the idea of dying at home, or at least outside the hospital. He noticed that his friend never allowed himself "poor me" talk. He refused to see dying as anything but part of living. This touched Jason deeply at the time, and now reflecting on that experience may help him develop realistic attitudes, too.

Communities.   What communities of people are there through which clients might identify strategies for implementing their goals? Even though Jason has not been a regular churchgoer, he does know that the parish within which he resides has some resources for helping those in need. A brief investigation reveals that the parish has developed a relatively sophisticated approach to visiting the sick. Parishioners are carefully selected and then trained for this program. They visit sick people in the hospital, in hospices, and at home and provide a number of services.

Places.   Are there particular places that might help? Jason immediately thinks of Lourdes, the shrine to which Catholic believers flock with all sorts of human problems. He doesn't expect miracles, but he feels that he might experience life more deeply there. It's a bit wild, but why not a pilgrimage? He still has the time and enough money to do it.

Things.   What things exist that can help clients achieve their goals? Jason has read about the use of combinations of drugs to help stave off pain and the side effects of chemotherapy. He has heard that certain kinds of electric stimulation can ward off chronic pain. He explores all these possibilities with his doctor and even arranges for second and third opinions.

Organizations.   Are there any organizations that help people with this kind of problem? Jason knows that there are mutual-help groups composed of cancer patients. He has heard of one at the hospital and believes that there are others in the community. He learns more about hospices for those terminally ill with cancer.

Programs.   Are there any ready-made programs for people in the client's position? A hospice for the terminally ill has just been established in the city. They have three programs. One helps people who are terminally ill stay in the community as long as they can. A second makes provision for part-time residents. The third provides a residential program for those who can spend little or no time in the community. The goals of these programs are practically the same as Jason's.

WHAT SKILLS DO CLIENTS NEED TO
GET WHAT THEY WANT?

People often get into trouble or fail to get out of it because they lack the needed life skills or coping skills and do not know how to mobilize resources, both internal and environmental, to cope with problem situations. If this is the case, then helping clients find ways of learning the life skills they need to cope more effectively is an important broad strategy. Indeed, the use of skills training as part of therapy — what Carkhuff (1971) called "training as treatment" — might be an essential for some clients. This is not a question of across-the-board training, but training that focuses on specific skills needed by clients to accomplish their goals. A constant question throughout the helping process should be what kinds of skills does this client need to get where he or she wants to go? Consider the following case.

> Darcy and Andreas fell in love. They married and enjoyed a relatively trouble-free honeymoon period of about two years or so. Eventually, however, the problems that inevitably arise from living together in such intimacy asserted themselves. They found, for instance, that they counted too heavily on positive feelings for each other and now, in their absence, could not communicate about finances, sex, and values. They lacked certain critical interpersonal communication skills. Furthermore, they lacked understanding of each other's developmental needs. Andreas had little working knowledge of the developmental demands of a 20-year-old woman; Darcy had little working knowledge of the kinds of cultural blueprints that were operative in the lifestyle of her 26-year-old husband. The relationship began to deteriorate. Since they had few problem-solving skills, they didn't know how to handle their situation.

In the case of this young couple, it seems reasonable to assume that helping will necessarily include both education and training as essential elements of an action plan.

One marriage counselor I know does marriage counseling in groups, usually of four couples. Training in communication skills is part of the process. He separates men from women and trains them in attending, listening, and empathy. For practice he begins by pairing a woman with a woman and a man with a man. Next he has a man and a woman — but not spouses — practice the skills together. Finally spouses are paired and are expected to use the skills they have learned to engage in problem solving with each other. Since he uses the helping model outlined in this book, his form of marriage counseling also includes, both directly and indirectly, training in basic problem-management skills. The helping model gives his clients a problem-management language they can use to talk to and help one another.

LINKING STRATEGY FORMULATION TO ACTION

Each of the steps of the helping process can and should stimulate action on the part of the client; this is especially true of Step III-A. Many clients, once they begin to see what they can do to get what they want, begin acting immediately. They don't need a formal plan. Here are two examples of clients who were helped to identify strategies for implementing their goals and to act on them.

> Jeff had been in the army for about ten months. He found himself overworked and, perhaps not paradoxically, bored. He had a couple of sessions with one of the educational counselors on the base. During these sessions Jeff began to see quite clearly that not having a high school diploma was working against him. The counselor mentioned that he could finish high school while in the army.

Jeff realized that this possibility had been pointed out to him during the orientation talks, but he hadn't paid any attention to it. He had joined the army because he wasn't interested in school and couldn't find a job. Now he decided that he would get a high school diploma as soon as possible.

Jeff obtained the authorization needed from his company commander to go to school. He found out what courses he needed and enrolled in time for the next school session. It didn't take him long to finish. Once he received his high school degree, he felt better about himself and found that opportunities for more interesting jobs opened up for him in the army. Achieving his goal of getting a high school degree helped him manage the problem situation.

Jeff was one of those fortunate ones who, with a little help, quickly set a goal (the "what") and identify and implement the strategies (the "how") to accomplish it. Note, too, that his goal of getting a diploma was also a means to other goals — feeling good about himself and getting better jobs in the army.

Grace's road to problem management was quite different from Jeff's. She needed much more help.

As long as she could remember, Grace had been a fearful person. She was especially afraid of being rejected and of being a failure. As a result, she had a rather impoverished social life and had held a series of jobs that were safe but boring. She became so depressed that she made a half-hearted attempt at suicide, probably more a cry of anguish and for help than a serious attempt to get rid of her problems by getting rid of herself.

During the resulting stay in the hospital, Grace had a few therapy sessions with one of the staff psychiatrists. The psychiatrist was supportive and helped her handle the guilt she felt because of the suicide attempt and the depression that had led to the attempt. Just talking to someone about things she usually kept to herself seemed to help. She began to see her depression as a case of learned helplessness. She saw quite clearly how she had let her choices be dictated by her fears. She also began to realize that she had a number of underused resources. For instance, she was intelligent and, though not good-looking, attractive in other ways. She had a fairly good sense of humor, though she seldom gave herself the opportunity to use it. She was also sensitive to others and basically caring.

After Grace was discharged from the hospital, she returned for a few outpatient sessions. She wanted to do something about her general fearfulness and her passivity, especially the passivity in her social life. A psychiatric social worker taught her relaxation and thought-control techniques that helped her reduce her anxiety. Once she felt less anxious, she was in a better position to do something about establishing some social relationships. With the social worker's help, she set goals of acquiring a couple of friends and becoming a member of some social group. However, she was at a loss as to how to proceed, since she thought that friendship and a fuller social life were things that should happen "naturally." She soon came to realize that many people had to work at acquiring a more satisfying social life, that for some people there was nothing automatic about it at all.

The social worker helped Grace identify various kinds of social groups that she might join. She was then helped to see which of these would best meet her needs without placing too much stress on her. She finally chose to join an arts and crafts group at a local YMCA. The group gave her an opportunity to begin developing some of her talents and to meet people without having to face demands for intimate social contact. It also gave her an opportunity to take a look at other, more socially oriented programs sponsored by the Y. In the arts and crafts program she met a couple of people she liked and who seemed to like her. She began having coffee with them once in a while and then an occasional dinner.

Grace still needed support and encouragement from her helper, but she was gradually becoming less anxious and feeling less isolated. Once in a while she would let her anxiety get the better of her. She would skip a meeting at the Y and then lie about having attended. However, as she began to let herself trust her helper more, she revealed this self-defeating game. The social worker helped her develop coping strategies for those times when anxiety seemed to be higher.

Grace's problems were more severe than Jeff's, and she did not have as many immediate resources. Therefore, she needed more time and more attention to develop goals and strategies for achieving them.

How effectively do I do the following?

- Use probes, prompts, and challenges to help clients identify possible strategies.
- Help clients engage in divergent thinking with respect to strategies.
- Help clients brainstorm as many ways as possible to accomplish their goals.
- Use some kind of framework in helping clients be more creative in identifying strategies.
- Help clients identify the skills they need to accomplish their goals.
- Help clients see the action implications of the strategies they identify.

CHAPTER 15

STEP III-B: BEST-FIT STRATEGIES — WHAT ACTIONS ARE BEST FOR THE CLIENT?

WHAT'S BEST FOR THE CLIENT?

HELPING CLIENTS CHOOSE STRATEGIES

STRATEGY SAMPLING

APPLYING THE BALANCE-SHEET METHOD TO STRATEGIES

Benefits of Choosing the Residental Program

Costs of Choosing the Residential Program

Using the Balance Sheet

THE SHADOW SIDE OF SELECTING STRATEGIES

LINKING STEP III-B TO ACTION

EVALUATION QUESTIONS FOR STEP III-B

WHAT'S BEST FOR THE CLIENT?

In the last two steps of Stage III, clients are in decision-making mode once more. After brainstorming strategies for accomplishing goals, they need to choose one or more strategies (a package) that best fit their situation and resources and turn them into a formal plan for constructive change. Whether these steps are done with the kind of formality outlined here is not the point. Counselors, who understand the technology of planning can add value by helping clients find ways of accomplishing goals (getting what they need and want) in a systematic and cost-effective way. Step III-B discusses ways of helping clients choose the strategies that are best for them. Step III-C deals with turning these strategies into some kind of step-by-step plan.

Some clients, once they are helped to develop a range of strategies to implement goals, move forward on their own; that is, they choose the best strategies, put together an action plan, and implement it. It is for the other clients—those who need help in choosing strategies that best fit their situation—that we add Step III-B (*see* Figure 15-1) to the helping process. It is useless to have clients brainstorm if they don't know what to do with all the action strategies they generate.

Consider the case of Charles, a man who was helped to discover two best-fit strategies for achieving emotional stability in his life against all odds.

> One morning, Charles, then 18 years old, woke up unable to speak or move. He was taken to a hospital and diagnosed as a catatonic schizophrenic. After repeated admissions to hospitals, where he underwent drug therapy and electroconvulsive therapy (ECT), his diagnosis was changed to paranoid schizophrenia. He was considered incurable.
>
> A quick overview of his earlier years suggests that much of his emotional distress was caused by unmanaged life problems and the lack of human support. He was separated from his mother for four years while he was young. They were reunited in a city new to both of them, and there he suffered a great deal of harassment at school because of his "ethnic" looks and accent. There was simply too much stress and change in his life. He protected himself by withdrawing. He was flooded with feelings of loss, fear, rage, and abandonment. Even small changes became intolerable. He had his catatonic attack one autumn when the time change occurred. It was the last straw.
>
> In the hospital Charles became convinced that he and many of his fellow patients could do something about their illnesses. They did not have to be victims of themselves or of the institutions designed to help them. Reflecting on his hospital stays and the drug and ECT treatments, he later said he found this "help" so disempowering that it was no wonder that he got crazier. Somehow, using his own inner resources, Charles managed to get out of the hospital. Eventually he got a job and married.
>
> One day, after a series of problems with his family and at work, Charles felt himself becoming agitated and thought he was choking to death. His doctor sent him to the hospital "for more treatment." There Charles had the good fortune to meet Sandra, a psychiatric social worker who was convinced that many of the hospital's patients were there because of lack of support before,

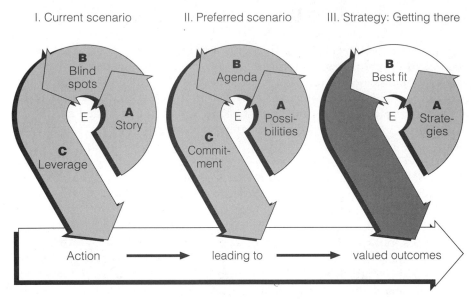

I. Current scenario   II. Preferred scenario   III. Strategy: Getting there

B Blind spots
A Story
E
C Leverage

B Agenda
A Possi-bilities
E
C Commit-ment

B Best fit
A Strate-gies
E

Action ⟶ leading to ⟶ valued outcomes

FIGURE 15-1
Step III-B: Choosing Best-Fit Strategies

during, and after their bouts of illness. She helped him see his need for social support, especially at times of stress. During the in-patient counseling groups that Sandra ran, she discovered that Charles had a knack of helping others. His broad goal was still emotional stability, and he wanted to do whatever was necessary to achieve it. Finding human support and helping others were his best strategies for achieving the stability he wanted. Sandra provided a great deal of support in the hospital and coached his wife on providing support at times of stress.

Outside, Charles started a self-help group for ex-patients like himself. He was a full-fledged participant in the group. Further, he turned this self-help group into a network of self-help groups for ex-patients.

This is an amazing example of a client who focused on one broad goal — emotional stability — and translated it into a number of immediate, practical goals. He then discovered two broad strategies — finding emotional support and helping others — for accomplishing these goals and translated these into practical applications. By doing all this, he found the emotional stability he sought.

HELPING CLIENTS CHOOSE STRATEGIES

The criteria for choosing goal-accomplishing strategies are somewhat like the criteria for choosing goals outlined in Step II-B. These criteria are reviewed briefly here through a number of examples. Strategies to achieve goals should be, like goals themselves, specific, robust, realistic, and in keeping with clients' values.

Specific strategies.   Strategies for achieving goals should be specific enough to drive behavior. In our example, Charles translated two broad strategies for achieving emotional stability — tapping into human support and helping others — into quite specific strategies — keeping in touch with Sandra, getting help from his wife, participating in a self-help group, starting a self-help group, and founding and running a self-help organization.

Substantive strategies.   Strategies are robust to the degree that they challenge the client's resources and, when implemented, actually achieve the goal. Charles's strategies were challenging, especially the strategy of starting and running a self-help organization. Contrast his case with Stacy's. Both ended up in the hospital, but the outcomes were completely different.

> Stacy was admitted to a mental hospital because she had been exhibiting bizarre behavior in her neighborhood. She dressed in a slovenly way and went around admonishing the residents of the community for their "sins." She was diagnosed as schizophrenic, simple type. She had been living alone for about five years, since the death of her husband. It seems that she had become more and more alienated from herself and others. In the hospital, medication helped control some of her symptoms. She stopped admonishing others and took reasonable care of herself, but she was still quite withdrawn. She was assigned to milieu therapy, a euphemism meaning that she followed the more or less benign routine of the hospital. She remained withdrawn and usually seemed moderately depressed. No therapeutic goals had been set, and the nonspecific program to which she was assigned was totally inadequate.

This program lacked bite. Sometimes courses of action are inadequate because the resources needed are not available. In this example, milieu therapy meant that the hospital lacked adequate therapeutic resources. Charles's strategies, on the other hand, proved to be powerful. They helped him stabilize and gave him a new perspective on life.

Realistic strategies.   If clients choose strategies that are beyond their resources, they are doing themselves in. Realistic strategies are within the resources of the client, under their control, and unencumbered by obstacles. Charles's strategies would have appeared unrealistic to most clients and helpers. But there is a point to be made. Just as we should help clients set stretch goals whenever possible, so we should not underestimate what clients are capable of doing. In the following case, the client moves from unrealistic to realistic strategies for getting what he wants.

> Desmond is in a halfway house after leaving a state mental hospital. From time to time he still has bouts of depression that incapacitate him for a few days. He wants to get a job because he thinks that a job will help him feel better about

himself, become more independent, and manage his depression better. He answers want ads in a rather random way and keeps getting turned down after being interviewed. He simply does not yet have the kinds of resources needed to put himself in a favorable light in job interviews. Moreover, he is not yet ready for a regular, full-time job. Desmond's counselor helps him explore some companies that have specific programs to help the disabled. These companies have found that some of their best workers come from the ranks of those rejected by others. After a few interviews Desmond gets a job that is within his capabilities.

There is, of course, a difference between realism and allowing clients to sell themselves short. Robust strategies that make clients stretch for a valued goal can be most rewarding.

Strategies in keeping with the client's values.   Make sure that the strategies chosen are consistent with the client's values.

Glenn was cited for battering his wife. Mandatory counseling was part of the suspended sentence he received. Outside the home he was even-tempered. At home he let himself go and took it out on his wife. The counselor discovered that Glenn and his wife were having sexual problems and that sexual frustration had a great deal to do with his anger. The counselor suggested that Glenn lower his frustration (and therefore control his anger) by engaging in sexual relations with other women. Glenn did this, but felt guilty and became depressed.

The counselor was wrong in suggesting a course of action without helping the client explore his values. He might have done so because he thought that Glenn's having sex with other women, if evil at all, would be less evil than his abusing his wife, but this is not the counselor's decision to make.

As another example, return to Charles's case. If Sandra had urged Charles to lower his expectations and adopt a safe lifestyle, she would have been suggesting strategies counter to his values.

STRATEGY SAMPLING

Some clients find it easier to choose strategies if they first sample some of the possibilities.

Karen, surprised by the number of program possibilities there were to achieve the goal of getting liquor out of her life, decided to sample some of them. She went to an open meeting of Alcoholics Anonymous, she attended a meeting of a women's lifestyle-issues group, she visited the hospital that had the residential treatment program, she joined up for a two-week trial physical-fitness program at a YMCA, and she had a couple of strategy meetings with her husband and children. None of this was done frantically, but it did occupy her energies and strengthened her resolve to do something about her alcoholism.

Of course, some clients could use strategy sampling as a way of putting off action. This was certainly not the case with Charles. His attending the meeting of a self-help group after leaving the hospital was a form of strategy sampling. He was impressed by the group, but he thought that he could start a group limited to ex-patients that would focus more directly on the kinds of issues ex-patients face.

APPLYING THE BALANCE-SHEET METHOD TO STRATEGIES

The balance-sheet method for decision making was reviewed briefly in Chapter 11. Here is an example of a client who used the balance-sheet method to assess the viability, not of a goal, but of strategies to achieve a goal. Karen's goal was to stop drinking. One possible strategy for accomplishing this goal was to spend a month as an inpatient at an alcoholic treatment center. This possibility appealed to her. However, since choosing this strategy would be a serious decision, the counselor, Joan, helped Karen use the balance sheet to weigh possible costs and benefits. Karen chose to consider the pluses and minuses for herself and for her husband and children. After Karen completed the balance sheets, Karen and Joan discussed Karen's findings.

Benefits of Choosing the Residential Program

Karen saw the benefits as follows:

- *For me*:   It would help me because it would be a dramatic sign that I want to do something to change my life. It's a clean break, as it were. It would also give me time just for myself. I'd get away from all my commitments to family, relatives, friends, and work. I see it as an opportunity to do some planning. I'd have to figure out how I would act as a sober person.
- *For significant others*:   I'm thinking mainly of my family here. It would give them a breather, a month without an alcoholic wife and mother around the house. I'm not saying that to put myself down. I think it would give them time to reassess family life and make some decisions about any changes they'd like to make. I think something dramatic like my going away would give them hope. They've had very little reason to hope for the last five years.

Acceptability of benefits.   Karen then considered the acceptability of the benefit she anticipated.

- *For me*:   I feel torn here. But looking at it just from the viewpoint of acceptability, I feel kind enough toward myself to give myself a month's time off. Also something in me longs for a new start in life. And it's not just time off. The program is a demanding one.

- *For significant others*:   I think that my family would have no problems in letting me take a month off. I'm sure that they'd see it as a positive step from which all of us would benefit.

Unacceptability of benefits.   Karen also considered the downside of those benefits.

- *For me*:   Going away for a month seems such a luxury, so self-indulgent. Also, even though taking such a dramatic step would give me an opportunity to change my current lifestyle, it would also place demands on me. My fear is that I would do fine while in the program, but that I would come out and fall on my face. I guess I'm saying it would give me another chance at life, but I have misgivings about having another chance. I need some help here.
- *For significant others*:   The kids are young enough to readjust to a new me. But I'm not sure how my husband would take this "benefit." He has more or less worked out a lifestyle that copes with my being drunk a lot. Though I have never left him and he has never left me, still I wonder whether he wants me back sober. Maybe this belongs under the "cost" part of this exercise. I need some help here. And, of course, I need to talk to my husband about all this. I also notice that some of my misgivings relate not to a residential program as such but to a return to a lifestyle free of alcohol. Doing this exercise helped me see that more clearly.

Costs of Choosing the Residential Program

Karen described the costs as follows:

- *For me*:   Well, there's the money. I don't mean the money just for the program, but I would be losing four weeks' wages. The major cost seems to be the commitment I have to make about a lifestyle change. And I know the residential program won't be all fun. I don't know exactly what they do there, but some of it must be demanding. Probably a lot of it.
- *For significant others*:   It's a private program, and it's going to cost the family a lot of money. The services I have been providing at home will be missing for a month. It could be that I'll learn things about myself that will make it harder to live with me — though living with a drinking spouse and mother is no joke. What if I come back more demanding of them — I mean, in good ways? I need to talk this through more thoroughly.

Acceptability of costs.   Karen then analyzed the costs of her decision and whether those costs would be acceptable.

- *For me*:   I have no problem at all with the money nor with whatever the residential program demands of me physically or psychologically. I'm willing to pay. What about the costs in terms of the demands the program will place on me for substantial lifestyle changes? Well, in principle I'm willing to pay what that costs. I need some help here.

■ *For significant others*: They will have to make financial sacrifices, but I have no reason to think that they would be unwilling. Still, I can't be making decisions for them. I see much more clearly the need to have a counseling session with my husband and children present. I think they're also willing to have a "new" person around the house, even if it means making adjustments and changing their lifestyle a bit. I want to check this out with them, but I think it would be helpful to do this with the counselor. I think they will be willing to come.

Unacceptability of costs.   She then reviewed what aspects of the costs might be unacceptable.

■ *For me*: I'm ready to change my lifestyle, but I hate to think that I will have to accept some dumb, dull life. I think I've been drinking, at least in part, to get away from dullness; I've been living in a fantasy world, a play world a lot of the time. A stupid way of doing it perhaps, but it's true. I have to do some life planning of some sort. I need some help here.

■ *For significant others*: It strikes me that my family might have problems with a sober me if it means that I will strike out in new directions. I wonder if they want the traditional homebody wife and mother. I don't think I could stand that. All this should come out in the meeting with the counselor.

Karen concluded: "All in all, it seems like the residential program is a good idea. There is something much more substantial about it than an outpatient program. But that's also what scares me."

Using the Balance Sheet

Karen's use of the balance sheet helped her make an initial program choice. It also enabled her to discover issues that she had not yet worked out completely. By using the balance sheet, she was able to return to the counselor with work to do rather than merely wondering what will happen next. This highlights the usefulness of exercises and other forms of structure that help clients take more responsibility for what happens both in the helping sessions and outside.

The balance sheet is not to be used with every client to work out the pros and cons of every course of action. There are more practical and flexible approaches to its use. You can use parts of the balance sheet and tailor them to the kinds of action strategies the client is exploring. In fact, one of the best uses of the balance sheet is not to use it directly at all. Keep it in the back of your mind whenever clients are making decisions. Use it as a filter to listen to clients. Then turn relevant parts of it into probes to help clients focus on issues they may be overlooking. "How will this decision affect the significant people in your life?" is a probe that originates in the balance sheet. "Is there any down side to that strategy?" might help a client who is being a bit too optimistic. Remember: this is a no-formula approach.

THE SHADOW SIDE OF
SELECTING STRATEGIES

The shadow side of decision making, discussed in Chapter 10, is certainly at work in clients' choosing strategies to implement goals. Goslin (1985) put it well.

> In defining a problem, people dislike thinking about unpleasant eventualities, have difficulty in assigning . . . values to alternative courses of action, have a tendency toward premature closure, overlook or undervalue long-range consequences, and are unduly influenced by the first formulation of the problem. In evaluating the consequences of alternatives, they attach extra weight to those risks that can be known with certainty. They are more subject to manipulation . . . when their own values are poorly thought through. . . . A major problem . . . for . . . individuals is knowing when to search for additional information relevant to decisions. (pp. 7–9)

In choosing courses of action, clients often fail to evaluate the risks involved and determine whether the risk is balanced by the probability of success. Gelatt, Varenhorst, and Carey (1972) suggested four ways in which clients may try to deal with the factors of risk and probability: wishful thinking, playing it safe, avoiding the worst, and achieving some kind of balance. The first three are often pursued without reflection and therefore lie in the shadows.

Wishful thinking.   In this case, clients choose a course of action that might (they hope) lead to the accomplishment of a goal regardless of risk, cost, or probability. For instance, Jenny wants her ex-husband to increase the amount of support he is paying for the children. She tries to accomplish this by constantly nagging him and trying to make him feel guilty. She does not consider the risk (he might get angry and stop giving her anything), the cost (she spends a lot of time and emotional energy arguing with him), or the probability of success (he does not react favorably to nagging). The wishful-thinking client operates blindly, engaging in some course of action without taking into account its usefulness. At its worst, this is a reckless approach. Clients who "work hard" and still "get nowhere" may be engaged in wishful thinking, persevering in using means they prefer but that are of doubtful efficacy.

Playing it safe.   In this case, clients choose only safe courses of action, ones that have little risk and a high degree of probability of producing at least limited success. For instance, Liam, a manager in his early 40s, is very dissatisfied with the way his boss treats him at work. His ideas are ignored, the authority he is supposed to have is preempted, and his boss has not responded to his attempts to discuss career development. His goals are to let his boss know about his dissatisfaction and to learn what his boss thinks about him and his career possibilities. However, he fails to bring these issues up when his boss is "out of sorts." Then, when things are going well, Liam doesn't want to "upset the apple cart." He drops hints about his dissatisfaction, even joking about it at times. He tells others of his dissatisfaction, in hopes that word will filter back to his boss. During formal appraisal sessions, he allows himself to be intimidated by his boss. However, in his own mind, he is doing whatever could be expected of a "reasonable" man. He does not know how safe he is playing it.

Avoiding the worst.   In this case, clients choose means that are likely to help them avoid the worst possible result. They try to minimize the maximum danger, often without identifying what that danger is. Crissy, dissatisfied with her marriage, sets a goal to be "more assertive." However, even though she has never said this either to herself or to her counselor, the maximum danger for her is to lose her partner. Therefore, her "assertiveness" is her usual pattern of compliance with some frills. For instance, every once in a while she tells her husband that she is going out with friends and will not be around for supper. He, without her knowing it, actually enjoys these breaks. At some level of her being, she realizes that her absences are not putting him under any pressure. She continues to be assertive in this way. But she never sits down with her husband in order to review where they stand with each other. This might be the beginning of the end.

Striking a balance.   In the ideal case, clients choose strategies for achieving goals that balance risks against the probability of success. This combination approach is the most difficult to apply, for it involves a great deal of analysis, including clarification of objectives, a solid knowledge of personal values, and the ability to rank a variety of action strategies according to one's values, plus the ability to predict results from a given course of action. Even more to the point, it demands challenging the shadow side of the problem, of goals chosen, and the ineffectual courses of action that have been chosen in the past. Therefore, some clients have neither the skill nor the will for this approach.

　　This brings up an interesting question. Is the job of the counselor to help clients achieve the "best" solution to their problems? The idealist within us says, "Of course." But this is not the reality of helping. In Chapter 3, one cluster of values was given the title "pragmatism." Of course we want to do the best job we can with our clients, but this does not mean that each stage and step of this helping model will be done perfectly. Helpers and clients both work with limited aspirations and limited resources. If clients engage in wishful thinking, play it safe, try to avoid the worst rather than seek the better, or waste a great deal of time and effort searching for the perfect course of action, counselors can challenge them to think through the consequences of their approach to decision making. In the end, however, skilled helpers do not make decisions for clients.

LINKING STEP III-B TO ACTION

Some clients are filled with great ideas for getting things done but never seem to do anything. They lack the discipline to evaluate their ideas, choose the best, and turn them into action. Often this kind of work seems too tedious to them, even though it is precisely what they need.

> Clint came away from the doctor feeling depressed. He was told that he was in the high-risk category for heart disease and that he needed to change his life-style. He was cynical, a man very quick to anger, a man who did not readily trust others. Venting his suspicions and hostility did not make them go away; it only intensified them. Therefore, one critical lifestyle change was to change this pattern and develop the ability to trust others. He developed three broad goals: reducing mistrust of others' motives; reducing the frequency and intensity of such emotions as rage, anger, and irritation; and learning how to treat others with consideration. Clint read through the strategies suggested to help people pursue these broad goals (see R. Williams, 1989):

- keeping a hostility log to discover the patterns of cynicism and irritation in one's life;
- finding someone to talk to about the problem, someone to trust;
- "thought stopping," catching oneself in the act of indulging in hostile thoughts or in thoughts that lead to hostile feelings;
- talking sense to oneself when tempted to put others down;
- developing empathic thought patterns, walking in the other's shoes;
- learning to laugh at one's silliness;
- using a variety of relaxation techniques, especially when negative thoughts come up;
- finding ways of practicing trust;
- developing active listening skills;
- substituting assertive for aggressive behavior;
- getting perspective, seeing each day as one's last; and
- practicing forgiving others without patronizing or condescending.

> Clint prided himself on his rationality (though his "rationality" was one of the things that got him in trouble). So, as he read down the list, he chose strategies that could form an "experiment," as he put it. He decided to talk to a counselor (for the sake of objectivity), keep a hostility log (data gathering), and use the tactics of thought stopping and talking sense to himself whenever he felt that he was letting others get under his skin. The counselor noted to himself that

none of these necessarily involved changing Clint's attitudes toward others. However, he did not challenge Clint at this point. His best bet was that through strategy sampling Clint would learn more about his problem, that he would find that it wont deeper than he thought. Clint set himself to his experiment with vigor.

Clint chose strategies that fit his values. The problem was that the values themselves needed reviewing. But Clint did act, and action gave him the opportunity to learn.

EVALUATION
QUESTIONS FOR STEP III-B

How well am I doing the following as I try to help clients choose actions that are best for them?

- Helping clients choose strategies that are clear and specific, that best fit their capabilities, that are linked to goals, that have power, and that are suited to clients' styles and values.
- Helping clients engage in and benefit from strategy sampling.
- Helping clients use the balance sheet as a way of choosing strategies by outlining the principal benefits and costs for self, others, and relevant social settings.
- Using the balance sheet in a flexible, client-centered way.
- Helping clients manage the shadow side of selecting courses of action.

STEP III-C: MAKING PLANS — HELPING CLIENTS DEVELOP ACTION PROGRAMS THAT WORK

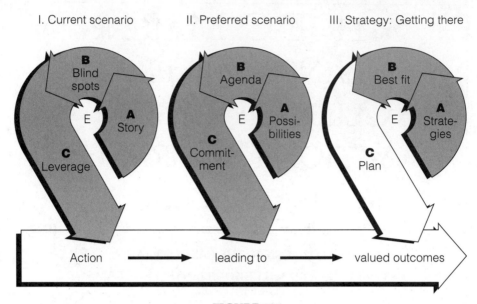

FIGURE 16-1
The Helping Model: Step III-C

We come to the last formal step of the helping process: making plans. This step requires figuring out what needs to be done first, second, third, and so forth, to make sure goals are accomplished. This step is illustrated in Figure 16-1.

A PLAN OF ACTION:
THE CASE OF FRANK

The lack of a plan — a clear step-by-step process in which strategies are used to achieve goals — keeps some clients mired in the psychopathology of the average. Consider the case of Frank, a vice-president of a large West Coast corporation.

> Frank was a go-getter. He was very astute about business and had risen quickly through the ranks. Vince, the president of the company, was in the process of working out his own retirement plans. From a business point of view, Frank was the heir apparent. But there was a glitch. The president was far more than a good manager; he was a leader. He had a vision of what the company should look like five to ten years down the line. Though tough, he related well to people. Frank was quite different. He was a hands-on manager. He was slow to delegate tasks to others, however competent they might be. He kept second-guessing others when he did delegate, he reversed their decisions in a way that made them feel put down, he was a poor listener, and he took a fairly short-term view of the business — "What were last week's figures like?" He was not a leader, but an operations man.

One day Vince sat down with Frank and told him that he was the heir apparent, but that his promotion would not be automatic. "Frank, if it were just a question of business acumen, you could take over today. But this job, at least in my mind, demands a leader." Vince went on to explain what he meant by a leader and to give Frank feedback on some of the things in his style that had to change.

Afterward Frank saw a consultant, someone he trusted who had given him similar feedback. They worked together for over a year, often over lunch and in hurried meetings early in the morning or late in the evening. Frank's aim was to become, within reason, the kind of leader Vince had described. Being very bright, he came up with some inventive strategies for moving in this direction. But he could never be pinned down to an overall program with specific milestones by which he could evaluate his progress. The consultant pushed him, but Frank was always "too busy" or would say that a formal program was not his "style." This was odd, since formal planning was one of his fortes in the business world.

Frank remained as astute as ever in his business dealings, and the business prospered. But he merely dabbled in the strategies meant to help him achieve his goal. At the end of two years, Vince retired after appointing someone else as president of the company.

Frank had two major blind spots that the consultant either could not or did not help him overcome. First, he thought the president's job was his, that business acumen alone would win out in the end. Second, he thought he could change his management style at the margins, when more substantial changes were called for. Frank had a lot of good ideas, but he never put them together into a coherent program for constructive change. Rather he jumped from one to another. He was more interested in "trying things" — that is, in activities and output — than in the outcome and impact on his managerial style. If the consultant had said, "Come on, Frank, I know you hate doing it, but let's map out a program and find ways to get you to stick to it," maybe things would have been different. It was not a question of changing Frank's entire personality, but of changing key patterns of behavior and doing so systematically.

Once they have determined what they need to do in order to get what they want, some clients have no difficulty deciding what needs to be done first, second, and so on. Others, though, need help at this stage. However, since some clients (and some helpers) fail to appreciate the power of a plan, it is useful to start by reviewing the advantages of planning.

HOW PLANS ADD VALUE TO CLIENTS' PROGRAMS FOR CHANGE

The focus here is on formal planning. The truth is that planning goes on throughout the helping process. Little plans, whether called such or not, are formulated and executed. Such plans are part of the little actions that clients engage in between sessions. Therefore, every session might have a covert planning dimension. A formal plan takes strategies for accomplishing goals, divides them into workable bits, puts the bits in order, and assigns a timetable for the accomplishment of each bit. Planning, when adapted to the needs of the client, has many advantages. Here are some of them.

- *Plans help clients develop needed discipline.*   Many clients get in trouble in the first place because they lack discipline. Planning places reasonable demands on clients to develop discipline.
- *Plans keep clients from being overwhelmed.*   Plans help clients see goals as doable. They keep the steps toward the accomplishment of a goal bite sized.
- *Formulating plans helps clients search for more useful ways of accomplishing goals — that is, even better strategies.*   When Mr. Johnson's wife and children began to formulate a plan for coping with their reactions to his alcoholism, they realized they had not been very inventive in looking for workable strategies. With the help of an Al-Anon self-help group member, they went back to the drawing board. Mr. Johnson's drinking had introduced a great deal of disorder into the family. Planning would help them restore order.
- *Plans provide an opportunity to evaluate the realism and adequacy of goals.*   When Walter began tracing out a plan to cope with the loss of his job and with a lawsuit filed against him by his former employer, he realized that his goals of getting a better job and of filing and winning a countersuit were unrealistic. He scaled back his expectations. Effective plans give clients a realistic picture of what they must actually do to achieve a goal. Only in drawing up the plan did Walter realize that he had set his goals unrealistically high.
- *Plans make clients aware of the resources they will need to implement their strategies.*   When Dora was helped by a counselor to formulate a plan to pull her life together after the disappearance of her young son, she realized that she couldn't do it alone. She had retreated from friends and even relatives, but now she knew she had to get back into community.
- *Formulating plans helps clients uncover unanticipated snags or obstacles to the accomplishment of goals.*   Only when Ernesto, the badly beaten gang member, started putting together a plan for a different lifestyle did he realize what an obstacle his lack of a high school diploma would be. He came to realize that he needed the equivalent of a diploma if he was to develop the kind of lifestyle he wanted.
- *Planning can help clients manage post-decisional depression.*   A variety of clients — one who had decided to give up smoking, another who had determined to leave an abusive husband, a couple who had agreed to end their separation and start reinventing their marriage, and a man who had a chancy operation for brain cancer — all had one thing in common. This was post-decisional depression, the angst that comes from wondering whether they had made the right decision. They were all helped to deal with this depression by getting on with the planning of the rest of their lives.

Formulating plans will not solve all of our clients' problems, but it is one way of making time an ally instead of an enemy. Many clients engage in aimless activity in their efforts to cope with problem situations. Plans help clients optimize the use of time.

SHAPING THE PLAN

Plans need shape to drive action. A formal plan identifies the activities or actions needed to accomplish a subgoal or goal, putting these activities into a logical order, and setting a time frame for the accomplishment of each key step. Therefore, there are three simple questions.

1. What concrete things need to be done to accomplish the goal or subgoal?
2. In what sequence should these be done? What should be done first? What second? What third?
3. What is the time frame? What should be done today? What tomorrow? What next month?

If a goal chosen by a client is complex or difficult, then it is useful to help the client establish subgoals as a way of moving step-by-step toward the ultimate goal. For instance, once Charles decided to start an organization of self-help groups composed of ex-patients from mental hospitals, there were a number of subgoals that needed to be accomplished before the organization would become a reality. For instance, he had to set up some kind of charter for the organization. "Charter in place" was one of the subgoals leading to his overall goal.

In general, the simpler the plan, the better, provided it helps clients achieve their goals. However, simplicity is not an end in itself. The question is not whether a plan or program is complicated but whether it is well-shaped. If complicated plans are broken down into subgoals and the strategies or activities needed to accomplish them, they are as capable of being achieved, if the time frame is realistic, as simpler ones. In schematic form, shaping looks like this:

Subprogram 1 leads to subgoal 1.
  Subprogram 2 leads to subgoal 2.
    Subprogram n (the last in the sequence) leads to the accomplishment of the ultimate goal.

The Case of Wanda

Take the case of Wanda, a client who set a number of goals in order to manage a complex problem situation. One of her goals was finding a job. The plan leading to this goal had a number of steps, each of which led to the accomplishment of a subgoal. The following subgoals were part of Wanda's job-finding program. They are stated as accomplishments (the past-participle approach).

> Subgoal 1: Résumé written.
> Subgoal 2: Kind of job wanted determined.
> Subgoal 3: Job possibilities canvassed.
> Subgoal 4: Best job prospects identified.
> Subgoal 5: Job interviews arranged.
> Subgoal 6: Job interviews completed.
> Subgoal 7: Offers evaluated.

The accomplishment of these subgoals led to the accomplishment of the overall goal of Wanda's plan: getting the kind of job she wanted.

Wanda also had to set up a step-by-step process to accomplish each of these subgoals. For instance, the process for accomplishing the subgoal "job possibilities canvassed" included such things as reading the Help Wanted section of the local papers, contacting friends or acquaintances who might have been able to offer jobs or provide leads, visiting employment agencies, reading the bulletin boards at school, and talking with someone in the job placement office.

Sometimes the sequencing of activities is important, sometimes not. In Wanda's case, it was important for her to draw up a generic résumé — one that she could tailor to each job possibility — before she began to canvass job opportunities. But when it came to using different methods for identifying available jobs, the sequence did not make any difference.

The Case of Harriet

Harriet, an undergraduate student at a small state college, wanted to become a helper. In her case, the sequence of activities was important. Although her college offered no formal program in applied psychology, Harriet identified several undergraduate courses that would help her toward her goal. One was called "Problem-Solving Approaches to Caring for Self and Others"; a second, "Effective Communication Skills"; a third, "Developmental Psychology: The Developmental Stages and Tasks of Late Adolescence and Early Adulthood." Harriet took the courses as they came up. Unfortunately, the first course was the one in problem-solving approaches, and it included brief one-on-one practice in the skills being learned. The further into the course Harriet went, the more she realized that she would have gotten more out of the course if she had taken the communication skills program first.

Harriet had also volunteered for the peer-helper program offered by the Center for Student Services. While those running the program were careful about whom they selected, they did not offer much training. Goodwill was supposed to take the volunteers a long way. Harriet realized that the developmental psychology course would have helped her in this program. Seeing that her activities needed to be better organized, she dropped out of the volunteer program and sat down with an adviser who had some background in psychology to come up with a reasonable plan. Harriet's is a clear example of a program for constructive change that should have been better shaped right from the beginning.

The Case of Frank, Revisited

Let's see what planning might have done for Frank, the vice-president who needed leadership skills. In this fantasy, Frank, like Scrooge, gets a second chance.

What does Frank need to do?   To become a leader, Frank decides to reset his style with his subordinates by involving them more in decision making. He wants to listen more, set work objectives through dialogue, ask subordinates for suggestions, and delegate more. He decides to visit subordinates to see what he can do to help them, coach them when they ask for advice, give them feedback, recognize their contributions, and reward them for achieving results beyond their objectives. Through his list, Frank is saying such things to himself as: "Listen more to those who report to you" "Coach those who need your advice." He believes that he does not have to get more specific than this. The package of things he will do with each subordinate will have to be tailored to the needs of each.

In what sequence should Frank do these things?   Frank decides that the first thing he will do is call in each subordinate and ask, "What do you need from me in order to get your job done? How can I add value to your work? And what management style on my part would help you most?" Their responses will help him decide which of the other activities will make sense with each subordinate. The second step is also clear. The planning cycle for the business is about to begin, and each manager needs to know what his or her objectives are. It is a perfect time to begin setting objectives. Frank therefore sends a memo to each of his direct reports, asking them to review the company's business plan and the plan for each of their functions, and to write down what they think their key managerial objectives for the coming year should be.

What is Frank's time frame?   Frank calls in each of his subordinates immediately to discuss what they need from him. He completes his objective-setting sessions with them within three weeks. He puts off further action on delegation until he gets a better reading on their objectives and what they need from him.

This gives you a rough idea of what a plan for Frank might have looked like and how it might have improved his abortive effort to change his managerial style. Notice that the plan does not spell out every last detail.

PLANNING IN THE REAL WORLD

Planning in the real world seldom looks like planning in the textbooks. Textbooks set out principles; helping sessions deal with human encounters. In

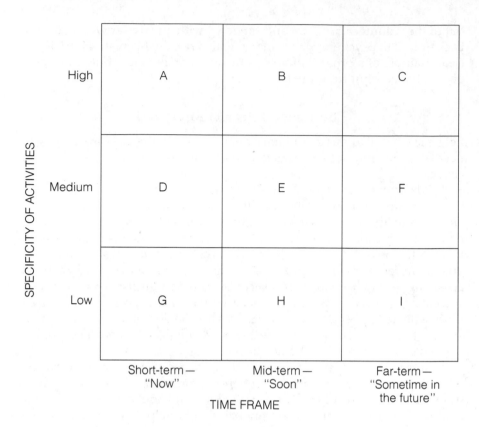

FIGURE 16-2
The Planning Grid

most cases, planning is probably done in bits and pieces throughout the helping process. "What do I need to do to get what I need and want? How can I get it done?" Questions like these are asked and answered either overtly or covertly at different times and in different ways. Kirschebaum (1985) challenged the notion that planning should always provide an exact blueprint of the specific activities to be engaged in, their sequencing, and the time frame. There are three questions:

1. How specific do the activities have to be?
2. How rigid does the order have to be?
3. How soon does each activity have to be carried out?

Kirschebaum suggested that, at least in some cases, being less specific and rigid in terms of activities, sequencing, and deadlines can "encourage people to pursue their goals by continually and flexibly choosing their activities" (p. 492). That is, flexibility in planning can help clients become more self-reliant and proactive. This is totally in keeping with our no-formula approach to helping.

Kirschebaum's points can be illustrated through the planning grid set forth in Figure 16-2. Putting everything in Box A — thus saying that every thing must be spelled out in great detail and everything must be done now — would be obsessive planning. Kirschebaum noted that rigid planning strategies can lead to frequent failure to achieve short-term goals. Then clients feel bad about themselves and give up trying. The other extreme — saying that "we will have to pull ourselves together one of these days — " is found in Box I. We need only look at our own experience to see how fatal that approach is. Overall, counselors should help clients establish the specificity and time frame that makes sense for them in their situations. There are no formulas; there are only client needs and common sense. Some things need to be done now; some, later. Some clients need more slack than others. Sometimes it helps to spell out the actions that need to be done in quite specific terms; at other times, it is only necessary to help clients outline them in broad terms and leave the rest to their own sound judgment. The art of counseling involves helping clients find the right specificity and time-frame tradeoffs.

Linking Step III-C to Action

As a counselor and as a consultant, I have learned one thing about plans: whether formulated by individuals or by organizations, many end up in a drawer someplace. Plans have a way of putting us face to face with the work we will have to do to accomplish our goals and touching off our resistance to change. As a friend once said, only half in jest, "I want to be happy, so I don't make plans."

Some clients are filled with great ideas for getting things done, but never seem to do anything. They lack the discipline to evaluate their ideas, choose the best, and then cast them into a step-by-step plan. This kind of work seems too tedious, even though it is precisely what they need. With these clients, you must be a planner — or rather, a consultant to their planning. Ask questions such as, "What are you going to do today? Tomorrow? The next day?" Such questions may sound very simple, but it is surprising how much leverage they can provide.

EVALUATION
QUESTIONS FOR STEP III-C

Helpers can ask themselves the following questions as they help clients formulate the kinds of plans that actually drive action.

- To what degree do I prize and practice planning in my own life?
- How quickly do I move to planning when I see that it is what clients need in order to manage problems and develop opportunities better?
- What do I do to help clients overcome inertia and resistance to planning?
- How effectively do I help clients formulate subgoals that lead to the accomplishment of overall preferred-scenario goals?
- How practical am I in helping clients identify the activities needed to accomplish subgoals, sequence these activities, and establish realistic time frames for them?
- How well do I adapt the specificity and detail of planning to the needs of each client?
- Even at this planning step, how easily do I move back and forth among the different stages and steps of the helping model as the need arises?
- How readily do clients actually move to action because of my work with them in planning?

CHAPTER 17

MAKING THINGS WORK—HELPING CLIENTS GET WHAT THEY WANT AND NEED

I. Current scenario　　　II. Preferred scenario　　　III. Strategy: Getting there

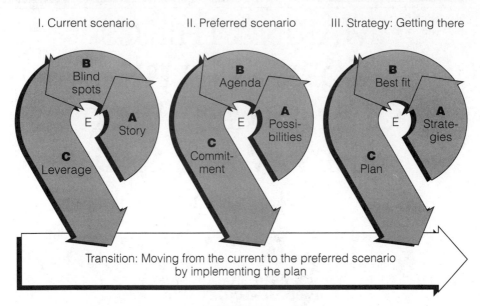

FIGURE 17-1
Transition: Moving from the Current to the Preferred Scenario

THE TRANSITION STATE

The action arrow of the helping model in Figure 17-1 constitutes the transition state in which clients move from the current to the preferred scenario by implementing strategies and plans (*see* Beckhard & Harris, 1987). However, as Marilyn Ferguson (1980) noted, clients often feel at risk during transition times. They have a feeling similar to that of letting go of one trapeze bar — familiar but dysfunctional patterns of behavior — and grabbing hold of the other bar — new and more productive patterns of behavior. Anticipating the terror of being in midair without support, some clients refuse to let go. Inertia wins. This is one reason that it is important to encourage clients to act, to begin the transition (at least in small ways) from the first session on. If clients start acting to change things right from the beginning, if each stage and step of the helping model is used as a driver of action, then they are less likely to be stymied by the actions called for by a more formal plan.

Since helpers don't follow their clients around in the clients' day-to-day lives, helpers cannot completely know how well their clients turn goals and strategies into problem-managing action. As indicated in Chapter 1, outcome research shows that helping does help, at least when viewed from a meta-analytic perspective. On the other hand, there is evidence that helpers know little about what their clients do once they leave the helping relationship (Lewis & Magoon, 1987). Perhaps it is safe to say that follow-up studies, when done, indicate that helping is helpful. In our terms, clients achieve their goals; they get what they want. But the studies deal with only a very small fraction of actual cases. The researchers tell us that these small samples add up to positive results. This clashes with the poor track record of discretionary change in human affairs.

Some clients, once they have a clear idea of what to do to handle a problem situation—whether or not they have a formal plan—indeed go ahead and do it. They need little or nothing in terms of further support and challenge from a helper. They either find the resources they need within themselves or get support and challenge from the significant others in the social settings of their lives. At the other end of the spectrum are clients who choose goals and come up with strategies for implementing them but who are, for whatever reason, stymied when it comes to action. Of course, other clients fall between these two extremes.

ENTROPY: THE SHADOW SIDE OF IMPLEMENTING CHANGE

Inertia, as discussed in Chapter 4, is the human tendency to put off problem-managing action. With respect to inertia, I often say to clients:

> The action program you've come up with seems to be a sound one. The main reason that sound action programs don't work, however, is that they are never tried. Don't be surprised if you feel reluctant to act or are tempted to put off the first steps. This is quite natural. Ask yourself what you can do to get by that initial barrier.

Since inertia comes right at the beginning, it is easier for counselors to help clients overcome it. However, overcoming initial inertia and launching a program for constructive change is one thing. Sustaining such a program is another.

Entropy is the tendency to give up action that has been initiated. Programs for constructive change, even those that start strong, often dwindle and disappear. All of us have experienced the problems involved in trying to implement programs. We make plans, and they seem realistic to us. We start the steps of a program with a good deal of enthusiasm. However, we soon run into tedium, obstacles, and complications. What seemed so easy in the planning stage now seems quite difficult. We become discouraged, flounder, recover, flounder again, and finally give up, offering ourselves rationalizations as to why we did not want to accomplish those goals anyway.

Phillips (1987) identified the "ubiquitous decay curve" in both helping and in medical-delivery situations (p. 650). Attrition, noncompliance, and relapse are the name of the game. A married couple trying to reinvent their marriage might eventually say to themselves, "We had no idea that it would be so hard to change ingrained ways of interacting with each other. Is it worth the effort?" Their motivation is on the wane. Wise helpers know that the decay curve is part of life and help clients deal with it. With respect to entropy, a helper might say: "Even sound action programs begun with the best of intentions tend to fall part over time, so don't be surprised when your initial enthusiasm seems to wane a bit. That's only natural. Rather, ask yourself what you need to do to keep yourself at the task."

Brownell, Marlett, Lichtenstein, and Wilson (1986) provided a useful caution. They drew a fine line between preparing clients for mistakes and giving them permission to make mistakes by implying that mistakes are inevitable. They also made a distinction between lapse and relapse. A slip or a mistake in an action program (a lapse) need not lead to giving up the program entirely (a relapse). Consider the following example. Graham has been trying to change what others see as his angry interpersonal style. Using a variety of self-monitoring and self-control techniques, he has made great progress in changing his style. On occasion he loses his temper, but never in any extreme way. His occasional lapse does not end up in relapse.

OVERCOMING ENTROPY: HELPING CLIENTS ENGAGE IN SUSTAINED GOAL-ACCOMPLISHING ACTION

Kanfer and Schefft (1988, p. 58) differentiated between decisional self-control and protracted self-control. In the former, a single choice terminates a conflict. For instance, a couple makes the decision to get a divorce. In the latter, continued resistance to temptation is required. For instance, a client has learned how to control her anger but must continue to be on guard. Most clients need both kinds of self-control to manage their lives better. A client's choice to give up alcohol completely (decisional self-control) needs to be complemented by the ability to handle inevitable longer-term temptations. Protracted self-control calls for a preventive mentality and a certain degree of street smarts. It is easier for the client who has given up alcohol to turn down an invitation to go to a bar in the first place than to sit in a bar all evening with friends and refrain from drinking.

Kirschebaum (1987) investigated factors involved in self-regulatory failure, and found that many things can contribute to giving up: low initial commitment to change, weak efficacy and outcome expectations, the use of self-punishment rather than self-reward, depressive thinking, failure to cope with emotional stress, lack of consistent self-monitoring, failure to use effective habit-change techniques, giving in to social pressure, failure to cope with initial relapse, and paying attention to the wrong things. This final factor is illustrated by the client who focuses on how the environment is doing him in rather than how he is failing to cope with the environment.

When clients come up with good plans of action, I sometimes say something like this:

> Let's play a little game. Now that you know what you want and have a plan to get there, let's assume that the probability of your implementing your plan in a

sustained way is zero. This is not meant to be a negative assumption, but rather a helpful challenge. What do you need to do to raise the odds that you will begin and sustain this course of action from very poor to moderately good? After thinking for a few moments, one client said, "Well, I'd have to do something about the way I give in to social pressure. If my friends do something, I do it. If they think what I'm doing is silly, I stop. I think that would be my biggest challenge." She then went on to figure out what she would have to do to change this habit.

HELPING CLIENTS BECOME EFFECTIVE TACTICIANS

In the implementation and follow-through phase of the helping process, action strategies need to be complemented by tactics and logistics. A *strategy* is a practical plan to accomplish some objective. *Tactics* is the art of adapting a plan to the immediate situation. This includes being able to change the plan on the spot in order to handle unforeseen complications. *Logistics* is the art of being able to provide the resources needed for the implementation of a plan when they are needed.

> Rebecca wanted to take an evening course in statistics during the summer so that the first semester of the following school year would be lighter. Having more time would enable her to act in one of the school plays, a high priority for her. But she didn't have the money to pay for the course and at the university she planned to attend, prepayment for summer courses was the rule. Rebecca had counted on paying for the course from her summer earnings, but she would not have the money until later. Consequently, she did some quick shopping and found that the same course was being offered by a community college not too far from where she had intended to go. Her tuition there was minimal, since she was a resident of the area the college served.

In this example, Rebecca kept to her overall plan (strategy). However, she adapted the plan to an unforeseen circumstance (tactics) by locating another resource (logistics).

Stein (1980) suggested that action programs need to be buttressed by "rachets": strategies and tactics designed to keep action programs from falling apart. Helping clients find the right set of rachets might be one of the best things a helper can do. There are many different ways in which counselors can help clients become effective tacticians as they grapple with the day-to-day problems and anxieties associated with pursuing goals. Kanfer and Schefft (1988, *see* pp. 105–114, 123–166, and 215–294) outlined a number of them.

- Modify the environment to make action easier.
- Simplify the action plan needed to accomplish the goal.
- Get rid of action-inhibiting habits.
- Find substitutions for unwanted behavior.

- Find mentors, models, and exemplars to emulate.
- Engage in self-monitoring.
- Build mini-successes into the program.
- Readjust the time frames of action programs.

Since many well-meaning and motivated clients are simply not good tacticians, counselors can use these and other methods to help clients fight off entropy and sustain action. A number of these techniques are discussed and illustrated in the following sections.

Help Clients Develop Contingency Plans

If counselors help clients brainstorm possibilities for a better future (goals) and strategies for achieving these goals (courses of action), then clients will have the raw materials, as it were, for developing contingency plans. Contingency plans help make clients more effective tacticians. The formulation of contingency plans is based on the fact that we live in an imperfect world. Goals may have to be finetuned or even changed. The same is true for strategies for accomplishing goals.

> Jason, who was dying of cancer, decided to become a resident in the hospice he had visited. The hospice had an entire program in place for helping patients like him die with dignity. However, although Jason had visited the hospice and had liked what he had seen, he could not be absolutely sure that being a resident there would work out. Therefore, with the help of the counselor, he settled on two other back-up possibilities. One was living at home with some outreach services from the hospice. The other was spending his last days in a smaller hospital in a nearby town. This last alternative would not be as convenient for his family, but he would feel more comfortable there, since he hated large hospitals.

Contingency plans are needed, especially when clients choose high-risk programs to achieve critical goals. Having back-up plans also helps clients develop more responsibility. If they see that a plan is not working, then they have to decide whether to try the contingency plan. Back-up plans need not be complicated. A counselor might merely ask: "If that doesn't work, then what will you do?" As in the case of Jason, clients can be helped to specify a contingency plan further once it is seen that the first choice is not working out.

Help Clients Be Forewarned and Forearmed

Forewarned is forearmed is the philosophy underlying force-field analysis (Lewin, 1969). This process, illustrated in Figure 17-2, is simply a review by the client of the major obstacles to and resources for the implementation of strategies and plans. Once clients see what their goals are and draw up plans to achieve these goals, they can be helped to discover what forces will

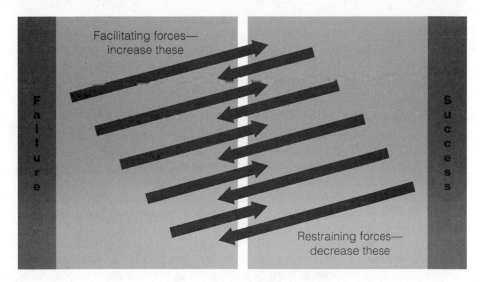

FIGURE 17-2
Force-Field Analysis

keep them from implementing their programs (restraining forces) and what forces will help them implement these programs (facilitating forces). This look into the future can save clients a great deal of grief down the line.

Restraining forces.   Restraining forces are the obstacles that might be encountered during the implementation process. The identification of possible obstacles to the implementation of a program helps make clients forewarned.

> Raul and Maria were a childless couple living in a large midwestern city. They had been married for about five years and had not been able to have children. They finally decided that they would like to adopt a child, and so they consulted a counselor familiar with adoptions. The counselor helped them work out a plan of action that included helping them examine their motivation and lifestyle, contacting an agency, and preparing themselves for an interview. After the plan of action had been worked out, Raul and Maria, with the help of the counselor, identified two restraining forces: the negative feelings that often arise on the part of prospective parents when they are being scrutinized by an adoption agency and the feelings of helplessness and frustration caused by the length of time and uncertainty involved in the process.

The assumption here is that if clients are aware of some of the wrinkles that can accompany any given course of action, they will be less disoriented when they encounter them. This part of the force-field analysis process is, at its best, a straightforward census of probable pitfalls rather than a self-defeating search for every possible thing that could go wrong.

Restraining forces can come from within the clients themselves, from others, from the social settings of their lives, and from larger environmental forces. Once a restraining force is identified, ways of coping with it can be identified. Sometimes simply being aware of a pitfall is enough to help clients mobilize their resources to handle it. At other times a more explicit coping strategy is needed. For instance, the counselor arranged a couple of role-playing sessions with Raul and Maria in which she assumed the role of the examiner at the adoption agency and took a hard line in her questioning. These rehearsals helped Raul and Maria stay calm during the actual interviews. The counselor also helped them locate a mutual-help group of parents working their way through the adoption process. The members of the group shared their hopes and frustrations and provided support for one another. In a word, Raul and Maria were trained to cope with the restraining forces they might encounter on the road toward their goal.

Facilitating forces.   In a more positive vein, force-field analysis can help clients identify important resources to be used in implementing action programs. Facilitating forces can be persons, places, or things. One client found the Bible and meditation facilitating forces.

> Nora found it extremely depressing to go to her weekly dialysis sessions. She knew that without them she would die—but she wondered whether life was worth living if she had to depend on a machine. The counselor helped her see how she was making life more difficult for herself by letting herself think such discouraging thoughts. He helped her learn how to think thoughts that would broaden her vision of the world instead of narrowing it down to herself, her pain and discomfort, and the machine. Nora was a religious person and found in the Bible a rich source of positive thinking. She initiated a new routine: The day before a visit to the clinic, she began to prepare herself psychologically by reading from the Bible. Then, as she traveled to the clinic and underwent treatment, she meditated slowly on what she had read.

In this case, the client substituted positive thinking, a facilitating force, for negative thinking, a restraining force.

The following steps may be used in helping clients use force-field analysis:

- Help clients list all the restraining forces that might make them discard an action program. Brainstorming is useful in this step, but care should be taken that it not turn into an exercise in despair.
- Help clients list all the facilitating forces that can help them persevere in implementing strategies and plans. This step deals with resources. It helps clients look at what they have going for themselves.
- Have clients underline the forces in each list (facilitating and restraining) that seem most critical with respect to carrying out the action plan. Clients cannot deal with every facilitating and every restraining force. Some are more important than others.

- Help clients identify ways of reducing significant restraining forces and of strengthening significant facilitating forces.

Obviously this process need not be used formally as outlined here. Rather, effective helpers have the force-field paradigm in their minds and use it to probe and challenge.

Help Clients Find Incentives and Rewards for Sustained Action

Clients avoid engaging in action programs when the incentives and rewards for not engaging in the program outweigh the incentives and rewards for doing so.

> Miguel kept saying that he wanted to leave his father's business and strike out on his own, especially since he and his father had heated arguments over how the business should be run. He earned an MBA in night school and talked about becoming a consultant to small, family-run businesses. A medium-sized consulting firm offered him a job. He accepted on the condition that he could finish up some work in the family business. But he always found "one more" project in the family business that needed his attention.

In his discussions with the counselor, Miguel learned some painful truths about himself. He prized comfort, and moving to another job would be uncomfortable, at least in the short term. He prized autonomy. His father let him do what he wanted. Almost any other job would mean less autonomy. The counselor helped Miguel focus on the rewards of moving on to another job: the chance to test his mettle, the opportunity to learn new skills, possibilities for advancement, and excellent salary prospects.

The incentive and rewards that help a client get going on a program of constructive change in the first place may not be the ones that keep the client going.

> Dwight, a man in his early 40s who was recovering from an accident at work that had left him partially paralyzed, was about to give up an arduous physical rehabilitation program. The counselor asked him to visit the children's ward. Dwight was shaken by the experience and amazed at the courage of many of the kids. He was especially struck by one teenager who was undergoing chemotherapy. "He seems so positive about everything," Dwight said. The counselor told him that the boy was tempted to give up, too. Dwight and the boy saw each other frequently. Dwight put up with the pain. The boy hung in there. Three months later the boy died. Dwight's response, besides grief, was, "I can't give up now; that would really be letting him down."

Dwight's partnership with the teenager proved to be an excellent incentive.

Activities that are not rewarded tend to lose their vigor, decrease, and even disappear over time. This process is called extinction.

Luigi had been in and out of mental hospitals a number of times. He discovered that one of the best ways of staying out was to use some of his excess energy to help others. He had not returned to the hospital once during the three years he worked at a soup kitchen. However, finding himself becoming more and more manic over the previous six months and fearing that he would be rehospitalized, he sought the help of a counselor.

Luigi's discussions with the counselor led to some interesting findings. They discovered that, in the beginning, Luigi had worked at the soup kitchen because he wanted to; now he was working there because he thought he should. He felt guilty about leaving and also thought that doing so would lead to a relapse. In sum, he had not lost his interest in helping others, but his current work was no longer interesting or challenging. As a result of his sessions with the counselor, Luigi began to work for a group that provided housing for the homeless and elderly. He poured his energy into his new work and no longer felt manic.

The lesson here is that incentives cannot be put in place and then be taken for granted. They need tending.

Help Clients Acquire the Skills They Need to Sustain Action

Sometimes clients do not act because they lack the working knowledge and skills required by strategies and action programs. If this is the case, then education and training can link planning to action. The principle is clear: either help clients devise action programs within the limitations of their current knowledge and skills or help them acquire the working knowledge and skills needed for more ambitious programs. The helper may be the trainer or may know where training can be obtained.

Ken, a gay man being seen by a counselor, was engaging in high-risk sexual activities even though he had determined to stop. Luckily, he was still HIV-negative. A helper referred Ken to a small-group education and training program based on the work of Kelly, Lawrence, Hood, and Brasfield (1989). Ken enrolled in this four-part program. In the first part, he learned about AIDS, HIV infection, how the virus is transmitted, and safer sex practices. In part two, using situations in which they had engaged in high-risk activities, group members learned how to analyze risky encounters in terms of such things as mood, emotional need, setting, and substance use. They considered these with a view to improving self-control and behavioral self-management. Part three involved assertiveness training, including how to get commitment from a partner to low-risk activities, how to resist demands for high-risk activities, and how to decline an immediate sexual proposition. In part four, group members learned the skills of establishing mutually supportive social relationships, social dating, and involvement in health-conscious activities. With the help of the program, Ken practically eliminated high-risk sexual activities.

It is important that the skills taught be related to the client's problems and opportunities. One advantage of training done in groups is social facilitation; group members provide support and challenge for one another as they put their newly learned skills to use.

Since clients do most of their problem-managing work outside the helping sessions, training them in self regulation skills (Watson & Tharp, 1993) can be most helpful. Indeed, if people were routinely trained in self-regulation skills early in life, fewer would be in need of helpers in the first place.

Help Clients Develop Action-Based Contracts

Earlier we discussed self-contracts as a way of helping clients commit themselves to what they want, to their goals. Self-contracts are also useful in helping them initiate and sustain problem-managing action. Feller (1984) developed the job-search agreement to help job seekers persist in their search. The following agreement requires clients to commit themselves to job-seeking behavior and to sound psychological practices that promote the right mentality for job-seeking behavior. Clients are to respond "True" to each of the following statements and then act on these truths.

I agree that no matter how many times I enter the job market, or the level of skills, experiences, or academic success I have, the following appear TRUE:

1. It takes only one YES to get a job; the number of NOs does not affect my next interview.
2. The open market lists about 20% of the jobs presently open to me.
3. About 80% of the job openings are located by talking to people.
4. The more people who know my skills and know that I'm looking for a job, the more I increase the probability that they'll tell me about a job lead.
5. The more specifically I can tell people about the problems I can solve or outcomes I can attain, rather than describe the jobs I've had, the more jobs they may think I qualify for.

I agree that regardless of how much I need a job, the following appear TRUE:

6. If I cut expenses and do more things for myself, I reduce my money problems.
7. The more I remain positive, the more people will be interested in me and my job skills.
8. If I relax and exercise daily, my attitude and health will appear attractive to potential employers.
9. The more I do positive things and the more I talk with enthusiastic people, the more I will gain the attention of new contacts and potential employers.
10. Even if things don't go as I would like them to, I choose my own thoughts, feelings, and behaviors each day.

Some contracts specify precisely what clients are to do and indicate rewards for success and sanctions for failures. Self-contracts are especially helpful for more difficult aspects of action programs; they help focus clients' energies.

In the following example, several parties had to commit themselves to the provisions of the contract.

> A boy in the seventh grade was causing a great deal of disturbance by his outbreaks in class. The usual kinds of punishment did not seem to work. After the teacher discussed the situation with the school counselor, the counselor called a meeting of all the stakeholders — the boy, his parents, the teacher, and the principal. The counselor offered a simple contract. When the boy disrupted the class, one and only one thing would happen: he would go home. Once the teacher indicated that the boy's behavior was disruptive, he was to go to the principal's office and check out without receiving any kind of lecture. He was to go immediately home and check in with whichever parent was at home, again without receiving any further punishment. The next day he was to return to school. All agreed to the contract, though both principal and parents said they would find it difficult not to add to the punishment.
>
> The first month the boy spent a fair number of days or partial days at home. The second month, however, he missed only two partial days, and the third month only one. The truth is that he really wanted to be in school with his classmates. That's where the action was. And so he paid the price of self-control in order to get what he wanted. The contract proved to be an effective tool of behavioral change.

The counselor had suspected that the boy found socializing with his classmates rewarding and being banished punishing.

GETTING ALONG WITHOUT A HELPER: SOCIAL SUPPORT AND CHALLENGE

In most cases, helping is a short-term process. Helpers use the skills and methods outlined in this book selectively to help clients deal with issues that are fairly focused. In the end, many clients (if not most) must make a go of it, empowered by the help received in their helping sessions. There are a number of possible scenarios at the implementation stage:

- Clients take the responsibility for implementation on their own and run with it.
- Clients continue to see a helper regularly in the implementation phase.
- Clients see a helper occasionally, either on demand or in scheduled stop-and-check sessions.
- The formal sessions end and it is up to the client to continue to pursue goals identified in the helping sessions or to maintain gains made there.

In all cases, clients benefit from learning what they need to do to persevere in their programs of constructive change. The methods and skills that help make clients more effective tacticians are part of this self-responsibility package.

However, many clients could benefit from some kind of ongoing helping process even after formal sessions have terminated. There are two distinct possibilities. First, clients can be helped to find some kind of appropriate self-help group to provide ongoing support and challenge. Second, clients can be helped to develop some kind of support system to continue the work started in the helping sessions. The self-help movement was reviewed briefly in Chapter 4; a few words about social support are in order here.

Social Support Systems

Clients' social networks are a two-edged sword. They can be obstacles to constructive change—and they can provide needed support (Bankoff & Howard, 1992). Clients often find counseling goals attractive and threatening at the same time. Since adherence to stressful decisions is painful, social support (Gibson & Brown, 1992; Sarason, Sarason, & Pierce, 1990) and challenge can help clients move to action, persevere in action programs, and maintain gains. Breier and Strauss (1984) discovered that supportive and challenging relationships can provide many benefits for clients, among them a forum for ventilation, reality testing, social support and approval, integration into a community, problem solving, and constancy. Their research focused on clients recovering from psychotic disorders, yet their findings have wider application.

Supportive relationships.   We have already discussed how necessary it is to help isolated clients develop social resources. These resources are critical when clients are acting on their own in the world. Protracted contact with the helper is usually not feasible, yet support remains essential. In clients' day-to-day environments, too little social support can be alienating; too much social support can be suffocating or may relieve clients of the burdens of self-responsibility. Individual differences, then, are important. The need for support is neither an admission of defeat nor an abdication of the principles of self-responsibility advocated in Chapter 3. It is a question of common sense. One problem with social support is that the kinds of supportive communication skills discussed earlier are not widely distributed in the population. People confuse sympathy with empathy and do not readily communicate the latter.

Challenging relationships.   It was suggested earlier that support without challenge can be hollow and that challenge without support can be abrasive. Ideally, the people in the lives of clients provide a judicious mixture of support and challenge. Or, put in a way that respects clients' self-responsibility, they help clients place reasonable demands on themselves.

Harry, a man in his early 50s, was suddenly stricken with a disease that called for immediate and drastic surgery. He came through the operation quite well, even getting out of the hospital in record time. For the first few weeks he seemed, within reason, to be his old self. However, he had problems with the drugs he had to take following the operation. He became quite sick and took on many of the mannerisms of a chronic invalid. Even after the right mix of drugs was found, he persisted in invalid-like behavior. Whereas right after the operation he had "walked tall," he now began to shuffle. He also talked constantly about his symptoms and generally used his "state" to excuse himself from normal activities.

At first Harry's friends were in a quandary. They realized the seriousness of the operation and tried to put themselves in his place. They provided all sorts of support. But gradually they realized that he was adopting a style that would alienate others and keep him out of the mainstream of life. Support was essential, but it was not enough. They used a variety of ways to challenge his behavior: mocking his "invalid" movements, engaging in serious one-to-one talks, turning a deaf ear to his discussion of symptoms, and routinely including him in their plans.

Harry did not always react graciously to his friends' challenges, but in his better moments he admitted that he was fortunate to have such friends.

Feedback from significant others.   In his book on human competence, Tom Gilbert (1978) claimed that "improved information has more potential than anything else I can think of for creating more competence in the day-to-day management of performance" (p. 175). Feedback is one way of providing support and challenge. If clients are to be successful in implementing their action plans, they need adequate information about how well they are performing. The purpose of feedback is not to pass judgment on the performance of clients, but rather to provide guidance, support, and challenge. There are two kinds of feedback.

1. Through *confirmatory* feedback, significant others such as helpers, relatives, friends, and colleagues let clients know that they are on course, that they are moving successfully through the steps of an action program toward a goal.
2. Through *corrective* feedback, significant others let clients know that they wandered off course and what they need to do to get back on course.

Corrective feedback, whether from helpers or people in the client's everyday life, should incorporate the following principles:

- It should be given in the spirit of caring.
- It should be coupled with confirmatory feedback.
- It should be brief and to the point.
- It should focus on clients' behaviors rather than on more elusive personality characteristics.
- It should be given in moderate doses. Overwhelming clients defeats the purpose of the entire exercise.

- The client should be invited to comment on the feedback and to expand on it.
- Clients should be helped to discover alternative ways of doing things.

One of the main problems with feedback is finding people in the client's day-to-day life who see the client in action enough to make their feedback meaningful, who care enough to give it, and who have the skills to provide it constructively.

Help Clients Fashion Social Support Systems

As I have mentioned earlier, many of the clients I see are, in one way or another, out of community or in community in unproductive ways. If they are out of community, over the course of the helping sessions, I try to help them find ways of reintegrating themselves—even if being out of community is not one of the presenting problems. Many simply have no community to speak of—at least no community of friends, acquaintances, or associates with whom they can share their problems and with whom they can take counsel.

This process pertains in one way to all clients. Eventually all clients have to make it on their own. Therefore, right from the beginning, I try to help them explore not only their primary concerns but also the quality of their social life. I do so because I know that they will need social support and challenge to accomplish the goals they will set along the way. In helping, all problem situations, opportunities, goals, and strategies should be related to the social system of the client. The kinds of questions that can pepper the helping interviews include:

- Who might help you do this?
- Who's going to challenge you when you want to give up?
- With whom can you share these kinds of concerns?
- Who's going to give you a pat on the back when you accomplish your goal?

If clients have no social systems, then helping them create their own social systems is important. This is not the same as trying to force clients to become more social than they want to be or to lose their independence.

CHOOSING NOT TO CHANGE

Some clients who seem to do well in analyzing problems, developing goals, and even identifying reasonable strategies and plans end up by saying—in effect, if not directly—something like this:

> I've explored my problems and understand why things are going wrong—that
> is, I understand myself and my behavior better—and I realize what I need to

do to change. But right now I don't want to pay the price called for by action. The price of more effective living is too high.

The question of human motivation seems almost as enigmatic now as it must have been at the dawning of the history of the human race. So often we seem to choose our own misery. Worse, we choose to stew in it rather than endure the relatively short-lived pain of behavioral change. Helpers can and should challenge clients to search for incentives and rewards for managing their lives more effectively. They should also help clients understand the consequences of not changing. But in the end it is the client's choice.

As indicated earlier, many client problems are coped with and managed, not solved. Consider the following case of a woman who certainly did not choose not to change. Her case is a good example of a no-formula approach to developing and implementing a program for constructive change.

> Vickey readily admits that she has never fully "conquered" her illness. Some 20 years ago, she was diagnosed as manic-depressive. The current picture looked something like this: she would spend about six weeks on a high, then the crash would come and for about six weeks she'd be in the pits. After that, she'd be normal for about eight weeks. This cycle meant many trips to the hospital. About 7 years into her illness, during a period in which she was in and out of the hospital, she made a decision, saying, "I'm not going back into the hospital again." With a little help, she put it more positively: "I will so have my life together and I will so manage it so as to never again be a candidate for a mental institution." This was her ultimate, non-negotiable goal.
>
> Starting with this declaration of intent, Vickey moved on, in terms of Step II-B, to spell out what she wanted:
>
> 1. She could channel the energy of her highs.
> 2. She would consistently manage or at least endure the depression and agony of her lows.
> 3. She would not disrupt the lives of others by her behavior.
> 4. She would not make self-defeating decisions when either high or low.
>
> With some help of a rather non-traditional counselor, Vickey began to do things to turn these goals into reality. She used her broad goals to provide direction for everything she did.
>
> Vickey moved back and forth among the stages and steps of the problem-managing process. She learned as much as she could about her illness, including about crisis times and how to deal with highs and lows. To manage her highs, she learned to channel her excess energy into useful—or at least non-destructive—activity. Some of her strategies for controlling her highs centered on the telephone. She set up her own phone-answering business and worked very hard to make it work. She knew instinctively that controlling her illness meant not just managing problems, but developing opportunities. During her free time she spent long hours on the phone with a host of friends, being careful not to overburden any one person. Phone marathons became part of her lifestyle. She made the point that a big phone bill was infinitely better than a stay in the hospital. She called the telephone her "safety valve."

At the time of her highs, she would do whatever she had to do to tire herself out and get some sleep, for she had learned that sleep was essential if she was to stay out of the hospital. This included working longer shifts at the business. She developed a cadre of supportive people, including her husband. She took special care not to overburden him. She made occasional use of a drop-in crisis center, but preferred avoiding any course of action that reminded her of the hospital.

It must be noted that the central issue in this case is the client's decision to stay out of the hospital. Her determination drove everything else. This case also exemplifies the spirit of action that ideally characterizes the implementation stage of the helping process. She did not let inertia get the best of her. She did not let entropy do her in. Here is a woman who, with occasional help from a counselor, took charge of her life. She set some goals and devised a set of simple strategies for accomplishing them. To this day, she has not gone back to the hospital. Some will say that she was not cured by this process. Her goal was not to be cured, but to lead as normal a life as possible in the real world. Some would say that her approach lacked elegance. It certainly did not lack results.

In Summary

Not all the methods outlined in this chapter will be used with all clients. Each client has to be seen in terms of his or her needs and capabilities. Adapt the process to the resources of clients while helping them stretch. With this caution in mind, here is a summary of the principal ideas related to helping clients make the transition to action.

- Before clients implement a formal course of action, use techniques such as force-field analysis to help them identify the major obstacles they will encounter and the major resources that will be available to them.
- Help clients develop clear pictures of the incentives that will help them stick to action programs.
- As clients implement programs, provide the kinds of support and challenge they need to give themselves as fully as possible to the work of constructive behavioral change. Even better, help clients identify sources of effective support and challenge in their everyday lives.
- Use the problem-management model itself to help clients cope with the kinds of problems that arise in any effort to translate plans into action. That is, help clients run these implementation problems through the model.
- Help clients evaluate the quality of their participation in the action program, the degree to which the program is helping them move toward their goals, and the degree to which the achievement of those goals is helping them manage the original problem situations.
- Finally, help clients decide to recycle any or all of the problem-management process, to focus on some other problem, or to terminate the helping relationship.

How well am I doing the following as I try to help this client make the transition to action?

- Understand how widespread both inertia and entropy are and how they are affecting clients.
- Generally help clients become effective tacticians.
- Help clients develop contingency plans.
- Use force-field analysis to help clients discover and manage obstacles to action.
- Use force-field analysis to help clients discover resources that will enable them to act.
- Help clients find the incentives and rewards they need to persevere in action.
- Help clients acquire the skills they need to act and sustain goal-accomplishing action.
- Prepare clients to get along without a helper.
- Help clients develop a social support and challenge system in their day-to-day lives.
- Face up to the fact that not every client wants to change.

BIBLIOGRAPHY

ACKOFF, R. (1974). *Redesigning the future.* New York: Wiley.

ALTMAIER, E. M., & Johnson, B. D. (1992). Health-related applications of counseling psychology: Toward health promotion and disease prevention across the life span. In S. D. Brown & R. W. Lent (Eds.), *Handbook of counseling psychology* (pp. 315–347). New York: Wiley.

AMERICAN PSYCHOLOGICAL ASSOCIATION. (1992). Ethical principles of psychologists and code of conduct. Washington, DC.

ANDERSON, J. R. (1993). Problem solving and learning. *American Psychologist, 48,* 35–44.

ANDREWS, J. D. W. (1991). *The active self in psychotherapy.* Boston: Allyn and Bacon.

ARGYRIS, C. (1957). *Personality and organization: The conflict between system and the individual.* New York: Harper & Row.

ARGYRIS, C. (1962). *Interpersonal competence and organizational effectiveness.* Homewood, IL: Dorsey Press.

ARGYRIS, C. (1964). *Integrating the individual and the organization.* New York: Wiley.

ARGYRIS, C. (1982). *Reasoning, learning, and action.* San Francisco: Jossey-Bass.

ATKINSON, D. R., Worthington, R. L., Dana, D. M., & Good, G.E. (1991). Etiology beliefs, preferences for counseling orientations, and counseling effectiveness. *Journal of Counseling Psychology, 38(3),* 258–264.

BAILEY, K. G., Wood, H. E., & Nava, G. R. (1992). What do clients want? Role of psychological kinship in professional helping. *Journal of Psychotherapy Integration, 2(2),* 125–147.

BALDWIN, B. A. (1980). Styles of crisis intervention: Toward a convergent model. *Journal of Professional Psychology, 11,* 113–120.

BANDURA, A. (1977). Self-efficacy: Toward a unifying theory of behavioral change. *Psychological Review, 84,* 191–215.

BANDURA, A. (1980). Gauging the relationship between self-efficacy judgment and action. *Cognitive Therapy and Research, 4,* 263–268.

BANDURA, A. (1982). Self-efficacy mechanism in human agency. *American Psychologist, 37,* 122–147.

BANDURA, A. (1986). *Social foundations of thought and action: A social cognitive theory.* Englewood Cliffs, NJ: Prentice-Hall.

BANDURA, A. (1989). Human agency in social cognitive theory. *American Psychologist, 44,* 1175–1184.

BANDURA, A. (1990). Foreword to E. A. Locke & G. P. Latham, *A theory of goal setting and task performance.* Englewood Cliffs, NJ: Prentice-Hall.

BANDURA, A. (1991). Human agency: The rhetoric and the reality. *American Psychologist, 46,* 157–161.

BANKOFF, E. A., & Howard, K. I. (1992). The social network of the psychotherapy patient and effective psychotherapeutic process. *Journal of Psychotherapy Integration, 2(4)*, 273–294.

BECKHARD, R., & Harris, R. T. (1987). *Organizational transitions: Managing complex change* (2nd ed.). Reading, MA: Addison-Wesley.

BEIER, E. G., & Young, D. M. (1984). *The silent language of psychotherapy: Social reinforcement of unconscious processes* (2nd ed.). New York: Aldine.

BENBENISHTY, R., & Schul, Y. (1987). Client-therapist congruence of expectations over the course of therapy. *British Journal of Clinical Psychology, 26(1)*, 17–24.

BENNETT, M. I., & Bennett, M. B. (1984). The uses of hopelessness. *American Journal of Psychiatry, 141*, 559–562.

BERENSON, B. G., & Mitchell, K. M. (1974). *Confrontation: For better or worse.* Amherst, MA: Human Resource Development Press.

BERGER, D. M. (1989). Developing the story in psychotherapy. *American Journal of Psychotherapy, 43(2)*, 248–259.

BERGIN, A. E. (1991). Values and religious issues in psychotherapy and mental health. *American Psychologist, 46(4)*, 394–403.

BERNARD, M. E. (Ed.). (1991). *Using rational-emotive therapy effectively.* New York: Plenum.

BERNARD, M. E., & DiGiuseppe, R. (Eds.). (1989). *Inside rational-emotive therapy.* San Diego: Academic Press.

BERNE, E. (1964). *Games people play.* New York: Grove Press.

BERNHEIM, K. F. (1989). Psychologists and the families of the severely mentally ill: The role of family consultation. *American Psychologist, 44*, 561–564.

BEUTLER, L. E., & Bergan, J. (1991). Values change in counseling and psychotherapy: A search for scientific credibility. *Journal of Counseling Psychology, 38(1)*, 16–24.

BEVAN, W. (1982). A sermon of sorts in three plus parts. *American Psychologist, 37*, 1303–1322.

BINDER, C. (1990). Closing the confidence gap. *Training,* September, 49–56.

BORGEN, F. H. (1992). Expanding scientific paradigms in counseling psychology. In S. D. Brown & R. W. Lent (Eds.), *Handbook of counseling psychology.* New York: Wiley.

BRAMMER, L. (1973). *The helping relationship: Process and skills.* Englewood Cliffs, NJ: Prentice-Hall.

BREIER, A., & Strauss, J. S. (1984). The role of social relationships in the recovery from psychotic disorders. *American Journal of Psychiatry, 141*, 949–955.

BROWNELL, K. D., Marlatt, G. A., Lichtenstein, E., & Wilson, G. T. (1986). Understanding and preventing relapse. *American Psychologist, 41*, 765–782.

BURNS, D. D., & Nolen-Hoeksema, S. (1992). Therapeutic empathy and recovery from depression in cognitive-behavioral therapy: A structural equation model. *Journal of Consulting and Clinical Psychology, 60(3)*, 441–449.

CARKHUFF, R. R. (1969a). *Helping and human relations: Vol. 1. Selection and training.* New York: Holt, Rinehart & Winston.

CARKHUFF, R. R. (1969b). *Helping and human relations: Vol. 2. Practice and research.* New York: Holt, Rinehart & Winston.

CARKHUFF, R. R. (1971). Training as a preferred mode of treatment. *Journal of Counseling Psychology, 18*, 123–131.

CARKHUFF, R. R. (1985). *PPD: Productive program development.* Amherst, MA: Human Resource Development Press.

CARKHUFF, R. R. (1987). *The art of helping* (6th ed.). Amherst, MA: Human Resource Development Press.

CARKHUFF, R. R., & Anthony, W. A. (1979). *The skills of helping: An introduction to counseling.* Amherst, MA: Human Resource Development Press.

CLARK, A. J. (1991). The identification and modification of defense mechanisms in counseling. *Journal of Counseling and Development, 69,* 231–236.

COLE, H. P., & Sarnoff, D. (1980). Creativity and counseling. *Personnel and Guidance Journal, 59,* 140–146.

COVEY, S. R. (1989). *The seven habits of highly effective people.* New York: Simon & Schuster.

COWEN, E. L. (1982). Help is where you find it. *American Psychologist, 37,* 385–395.

CUMMINGS, N. A. (1979). Turning bread into stones: Our modern antimiracle. *American Psychologist, 34,* 1119–1129.

DEFFENBACHER, J. L. (1985). A cognitive-behavioral response and a modest proposal. *Counseling Psychologist, 13,* 261–269.

DEUTSCH, M. (1954). Field theory in social psychology. In G. Lindzey (Ed.), *The handbook of social psychology* (Vol. 1). Cambridge, MA: Addison-Wesley.

DIMOND, R. E., Havens, R. A., & Jones, A. C. (1978). A conceptual framework for the practice of prescriptive eclecticism in psychotherapy. *American Psychologist, 33,* 239–248.

DORN, F. J. (1984). *Counseling as applied social psychology: An introduction to the social influence model.* Springfield, IL: Charles C Thomas.

DORN, F. J. (Ed.). (1986). *The social influence process in counseling and psychotherapy.* Springfield, IL: Charles C Thomas.

DRISCOLL, R. (1984). *Pragmatic psychotherapy.* New York: Van Nostrand Reinhold.

EDWARDS, J., Tindale, R. S., Heath, L., & Posavac, E. J. (Eds.). (1990). *Social influence processes and prevention.* New York: Plenum.

EGAN, G. (1970). *Encounter: Group processes for interpersonal growth.* Pacific Grove, CA: Brooks/Cole.

EGAN, G. (1984). People in systems: A comprehensive model for psychosocial education and training. In D. Larson (Ed.), *Teaching psychological skills: Models for giving psychology away.* Pacific Grove, CA: Brooks/Cole.

EGAN, G., & Cowan, M. A. (1979). *People in systems: A model for development in the human-service professions and education.* Pacific Grove, CA: Brooks/Cole.

ELIAS, M. J., & Clabby, J. F. (1992). *Building social problem-solving skills.* San Francisco: Jossey-Bass.

ELLIOTT, R. (1985). Helpful and nonhelpful events in brief counseling interviews: An empirical taxonomy. *Journal of Counseling Psychology, 32,* 307–322.

ELLIS, A. (1984). Must most psychotherapists remain as incompetent as they are now? In J. Hariman (Ed.), *Does psychotherapy really help people?* Springfield, IL.: Charles C Thomas.

ELLIS, A. (1985). *Overcoming resistance: Rational-emotive therapy with difficult clients.* New York: Springer.

ELLIS, A. (1987a). The evolution of rational-emotive therapy (RET) and cognitive behavior therapy (CBT). In J. K. Zeig (Ed.), *The evolution of psychotherapy.* New York: Brunner/Mazel.

ELLIS, A. (1987b). Integrative developments in rational-emotive therapy (RET). *Journal of Integrative and Eclectic Psychotherapy, 6,* 470–479.

ELLIS, A., & DRYDEN, W. (1987). *The practice of rational-emotive therapy.* New York: Springer.

FARRELLY, F., & Brandsma, J. (1974). *Provocative therapy.* Cupertino, CA: Meta Publications.

FASSINGER, R. E., & Schlossberg, N. K. (1992). Understanding the adult years: Perspectives and implications. In S. D. Brown & R. W. Lent (Eds.), *Handbook of counseling psychology.* New York: Wiley.

FELLER, R. (1984). *Job-search agreements.* Monograph, Colorado State University, Fort Collins.

FERGUSON, M. (1980). *The aquarian conspiracy: Personal and social transformation in the 1980s.* Los Angeles: J. P. Tarcher.

FERGUSON, T. (1987, January–February). Agreements with yourself. *Medical Self-Care,* 44–47.

FESTINGER, S. (1957). *A theory of cognitive dissonance.* New York: Harper & Row.

FISHER, R., & Ury, W. (1981). *Getting to yes: Negotiating agreement without giving in.* Boston: Houghton Mifflin.

FRANCES, A., Clarkin, J., & Perry, S. (1984). *Differential therapeutics in psychiatry.* New York: Brunner/Mazel.

FRANK, J. D. (1973). *Persuasion and healing* (2nd ed.). Baltimore: Johns Hopkins University Press.

FREEDHEIM, D. K., (Ed.). (1992). *History of psychotherapy: A century of change.* Washington, D.C.: American Psychological Association.

FREEMAN, A., & Dattilio, F. M. (Eds.). (1992). *Comprehensive casebook of cognitive therapy.* New York: Plenum.

FREIRE, P. (1970). *Pedagogy of the oppressed.* New York: Seabury.

FREMONT, S. K., & Anderson, W. (1986). What client behaviors make counselors angry? An exploratory study. *Journal of Counseling and Development, 65,* 67–70.

FREMONT, S. K., & Anderson, W. (1988). Investigation of factors involved with therapists' annoyance with clients. *Professional Psychology: Theory and Research, 19,* 330–335.

FRETZ, B. R., & Simon, N. P. (1992). Professional issues in counseling psychology: Continuity, change, and challenge. In S. D. Brown & R. W. Lent (Eds.), *Handbook of counseling psychology.* New York: Wiley.

FRIEDLANDER, M. L., & Schwartz, G. S. (1985). Toward a theory of strategic self-presentation in counseling and psychotherapy. *Journal of Counseling Psychology, 32,* 483–501.

GALASSI, J. P., & Bruch, M. A. (1992). Counseling with social interaction problems: Assertion and social anxiety. In S. D. Brown & R. W. Lent (Eds.), *Handbook of counseling psychology.* New York: Wiley.

GARTNER, A., & Riessman, F. (1977). *Self-help in the human services.* San Francisco: Jossey-Bass.

GARTNER, A., & Riessman, F. (Eds.). (1984). *The self-help revolution.* New York: Human Sciences Press.

GASTON, L. (1990). The concept of the alliance and its role in psychotherapy: Theoretical and empirical considerations. *Psychotherapy, 27,* 143–153.

GAZDA, G. M. (1973). *Human relations development: A manual for educators.* Boston: Allyn & Bacon.

GELATT, H. B. (1989). Positive uncertainty: A new decision-making framework for counseling. *Journal of Counseling Psychology, 36,* 252–256.

GELATT, H. B., Varenhorst, B., & Carey, R. (1972). *Deciding: A leader's guide.* Princeton, NJ: College Entrance Examination Board.

GEORGES, J. C. (1988). Why soft-skills training doesn't take. *Training,* April, 40–47.

GERSTLEY, L., McLellan, A. T., Alterman, A. I., Woody, G. E., et al. (1989). Ability to form an alliance with the therapist: A possible maker of prognosis for patients with antisocial personality disorder. *American Journal of Psychiatry, 146(4)*, 508–512.

GIBB, J. R. (1968). The counselor as a role-free person. In C. A. Parker (Ed.), *Counseling theories and counselor education*. Boston: Houghton Mifflin.

GIBB, J. R. (1978). *Trust: A new view of personal and organizational development*. Los Angeles: The Guild of Tutors Press.

GIBSON, J., & Brown, S. D. (1992). Counseling adults for life transitions. In S. D. Brown & R. W. Lent (Eds.), *Handbook of counseling psychology*. New York: Wiley.

GILBERT, L. A. (1992). Gender and counseling psychology: Current knowledge directions for research and social action. In S. D. Brown & R. W. Lent (Eds.), *Handbook of counseling psychology*. New York: Wiley.

GILBERT, T. F. (1978). *Human competence: Engineering worthy performance*. New York: McGraw-Hill.

GOLDFRIED, M. R., & Castonguay, L.G. (1992). The future of psychotherapy integration. *Psychotherapy, 29(1)*, 4–10.

GOLDSTEIN, A. P. (1980). Relationship-enhancement methods. In F. H. Kanfer & A. P. Goldstein (Eds.), *Helping people change: A textbook of methods* (2nd ed.). New York: Pergamon Press.

GOSLIN, D. A. (1985). Decision making and the social fabric. *Society, 22(2)*, 7–11.

GREENBERG, L. S. (1986). Change process research. *Journal of Consulting and Clinical Psychology, 54*, 4–9.

GRENCAVAGE, L. M., & Norcross, J. C. (1990). Where are the commonalities among the therapeutic common factors? *Professional Psychology: Research and Practice, 21(5)*, 372–378.

HALEY, J. (1976). *Problem solving therapy*. San Francisco: Jossey-Bass.

HALL, E. T. (1977). *Beyond Culture*. Garden City, NJ: Anchor Press.

HALLECK, S. L. (1988). Which patients are responsible for their illnesses? *American Journal of Psychotherapy, 42*, 338–353.

HANDELSMAN, M. M., & Galvin, M. D. (1988). Facilitating informed consent for outpatient psychotherapy: A suggested written format. *Professional Psychology: Research and Practice, 19(2)*, 223–225.

HARE-MUSTIN, R., & Marecek, J. (1986). Autonomy and gender: Some questions for therapists. *Psychotherapy, 23*, 205–212.

HARVARD MENTAL HEALTH LETTER. (1993a, March). Self-help groups — Part I.

HARVARD MENTAL HEALTH LETTER. (1993b, April). Self-help groups — Part II.

HEADLEE, R., & Kalogjera, I. J. (1988). The psychotherapy of choice. *American Journal of Psychotherapy, 42(4)*, 532–542.

HENDRICK, S. S. (1990). A client perspective on counselor disclosure. *Journal of Counseling & Development, 69*, 184–185.

HEPPNER, P. P. (1989). Identifying the complexities within clients' thinking and decision making. *Journal of Counseling Psychology, 36*, 257–259.

HEPPNER, P. P., & Claiborn, C. D. (1989). Social influence research in counseling: A review and critique [Monograph]. *Journal of Counseling Psychology, 36*, 365–387.

HEPPNER, P. P., & Frazier, P. A. (1992). Social psychological processes in psychotherapy: Extrapolating basic research to counseling psychology. In S. D. Brown & R. W. Lent (Eds.), *Handbook of counseling psychology*. New York: Wiley.

HESKETH, B., Shouksmith, G. & Kang, J. (1987). A case study and balance sheet approach to unemployment. *Journal of Counseling and Development, 66*, 175–179.

HILL, C. E., & Corbett, M. M. (1993). A perspective on the history of process and outcome research in counseling psychology. *Journal of Counseling Psychology*, *40(1)*, 3–24.

HILLS, M. D. (1984). *Improving the learning of parents' communication skills by providing for the discovery of personal meaning*. Doctoral dissertation, University of Victoria, British Columbia, Canada.

HORVATH, A. O., & Symonds, B.D. (1991). Relation between working alliance and outcome in psychotherapy: A meta-analysis. *Journal of Counseling Psychology*, *38*, 139–149.

HOWARD, G. S. (1991). Culture tales: A narrative approach to thinking, cross-cultural psychology, and psychotherapy. *American Psychologist, 46*, 187–197.

HOWARD, G. S., & Conway, C. G. (1986). Can there be an empirical science of volitional action? *American Psychologist, 41*, 1241–1251.

HOWARD, G. S., Curtin, T. D., & Johnson, A. J. (1991). Point estimation techniques in psychological research: Studies on the role of meaning in self-determined action. *Journal of Counseling Psychology, 38(2)*, 219–226.

HOWARD, G. S., & Myers, P. R. (1990). Predicting human behavior: Comparing idiographic, nomothetic, and agentic methodologies. *Journal of Counseling Psychology, 37(2)*, 227–233.

HOWELL, W. S. (1982). *The empathic communicator*. Belmont, CA: Wadsworth.

HUBER, C. H., & Baruth, L. G. (1989). *Rational-emotive family therapy*. New York: Springer.

HURVITZ, N. (1970). Peer self-help psychotherapy groups and their implication for psychotherapy. *Psychotherapy: Theory, Research, and Practice, 7*, 41–49.

HURVITZ, N. (1974). Similarities and differences between conventional psychotherapy and peer self-help psychotherapy groups. In P. S. Roman & H. M. Trice (Eds.), *The sociology of psychotherapy*. New York: Aronson.

IMBER, S. D. (1992). Then and now: Forty years in psychotherapy research. *Clinical Psychiatric Review, 12(2)*, 199–204.

ISHIYAMA, F. I. (1990). A Japanese perspective on client inaction: Removing attitudinal blocks through Morita therapy. *Journal of Counseling & Development, 68*, 566–570.

JACKSON, A. T., & Meadows, F. B., Jr. (1991). Getting to the bottom to understand the top. *Journal of Counseling & Development, 70*, 72–76.

JANIS, I. L. (1983). The role of social support in adherence to stressful decisions. *American Psychologist, 38*, 143–160.

JANIS, I. L., & Mann, L. (1977). *Decision making: A psychological analysis of conflict, choice, and commitment*. New York: Free Press.

JANOSIK, E. H. (Ed.). (1984). *Crisis counseling: A contemporary approach*. Belmont, CA: Wadsworth.

JEFFERY, R. W. (1989). Risk behaviors and health. *American Psychologist, 44(9)*, 1194–1202.

JENSEN, J. P., Bergin, A. E., & Greaves, D. W. (1990). The meaning of eclecticism: New survey and analysis of components. *Professional Psychology: Research and Practice, 21*, 124–130.

JONES, A. S., & Gelso, C. J. (1988). Differential effects of style of interpretation: Another look. *Journal of Counseling Psychology, 35(4)*, 363–369.

KAGAN, N. (1973). Can technology help us toward reliability in influencing human interaction? *Educational Technology, 13*, 44–51.

KAHN, M. (1990). *Between therapist and client*. New York: W. H. Freeman.

KAMINER, W. (1992). *I'm dysfunctional, you're dysfunctional*. Reading, MA: Addison-Wesley.

KANFER, F. H., & Schefft, B. K. (1988). *Guiding therapeutic change*. Champaign, IL: Research Press.

KARASU, T. B. (1986). The specificity versus nonspecificity dilemma: Toward identifying therapeutic change agents. *American Journal of Psychiatry, 143*, 687–695.

KAUFMAN, G. (1989). *The psychology of shame*. New York: Springer.

KAYE, H. (1992). *Decision power*. New York: Prentice Hall.

KELLY, J. A., St. Lawrence, J. S., Hood, H. V., & Brasfield, T. L. (1989). Behavioral intervention to reduce AIDS risk activities. *Journal of Consulting and Clinical Psychology, 57*, 60–67.

KENDALL, P. C. (1992). Healthy thinking. *Behavior Therapy, 23(1)*, 1–11.

KENDALL, P. C., Kipnis, D., & Otto-Salaj, L. (1992). When clients don't progress: Influences on and explanations for lack of therapeutic progress. *Cognitive Therapy & Research, 16(3)*, 269–281.

KERR, B. & Erb, C. (1991). Career counseling with academically talented students: Effects of a value-based intervention. *Journal of Counseling Psychology, 38(3)*, 309–314.

KIERULFF, S. (1988). Sheep in the midst of wolves: Person-responsibility therapy with criminals. *Professional Psychology: Research and Practice, 19*, 436–440.

KIRSCHEBAUM, D. S. (1985). Proximity and specificity of planning: A position paper. *Cognitive Therapy and Research, 9*, 489–506.

KIRSCHEBAUM, D. S. (1987). Self-regulatory failure: A review with clinical implications. *Clinical Psychological Review, 7*, 77–104.

KIVLIGHAN, D. M., Jr. (1990). Relation between counselors' use of intentions and clients' perception of working alliance. *Journal of Counseling Psychology, 37*, 27–32.

KIVLIGHAN, D. M., Jr., & Schmitz, P. J. (1992). Counselor technical activity in cases with improving work alliances and continuing-poor working alliances. *Journal of Counseling Psychology, 39*, 32–38.

KNAPP, M. L. (1978). *Nonverbal communication in human interaction* (2nd ed.). New York: Holt, Rinehart & Winston.

KOHUT, H. (1978). The psychoanalyst in the community of scholars. In P. H. Ornstein (Ed.), *The search for self: Selected writings of H. Kohut*. New York: International Universities Press.

KOTTLER, J. A. (1986). *On being a therapist*. San Francisco: Jossey-Bass.

KOTTLER, J. A. (1992). *Compassionate Therapy*. San Francisco: Jossey-Bass.

LANDRETH, G. L. (1984). Encountering Carl Rogers: His views on facilitating groups. *Personnel and Guidance Journal, 62*, 323–326.

LANGER, E. (1989). *Mindfulness*. Reading, MA: Addison-Wesley.

LATTIN, D. (1992, November 21). Challenging organized religion: Spiritual small-group movement. *San Francisco Chronicle*, pp., A1, A9.

LAZARUS, A. A. (1976). *Multimodal behavior therapy*. New York: Springer.

LAZARUS, A. A. (1981). *The practice of multimodal therapy*. New York: McGraw-Hill.

LAZARUS, A. A., Beutler, L. E., & Norcross, J. C. (1992). The future of technical eclecticism. *Psychotherapy, 29(1)*, 11–20.

LEAHEY, M., & Wallace, E. (1988). Strategic groups: One perspective on integrating strategic and group therapies. *Journal for Specialists in Group Work, 13*, 209–217.

LEE, C., & Richardson, B. (Eds.). (1991). *Multicultural issues in counseling: New approaches to diversity*. Alexandria, VA: American Association for Counseling and Development.

LEWIN, K. (1969). Quasi-stationary social equilibria and the problem of permanent change. In W. G. Bennis, K. D. Benne, & R. Chin (Eds.), *The planning of change*. New York: Holt, Rinehart & Winston.

LEWIS, J. D., & Magoon, T. M. (1987). Survey of college counseling centers' follow-up practices with former clients. *Professional Psychology: Research and Practice, 18*, 128–133.

LLEWELYN, S.P. (1988). Psychological therapy as viewed by clients and therapists. *British Journal of Clinical Psychology, 27(3)*, 223–237.

LOCKE, E. A., & Latham, G. P. (1984). *Goal setting: A motivational technique that works*. Englewood Cliffs, NJ: Prentice-Hall.

LOCKE, E. A., & Latham, G. P. (1990). *A theory of goal setting and task performance*. Englewood Cliffs, NJ: Prentice-Hall.

LOCKE, E. A., Shaw, K. N., Saari, L. M., & Latham, G. P. (1981). Goal setting and task performance: 1969–1980. *Psychological Bulletin, 90*, 125–152.

LONDON, P. (1986). *The modes and morals of psychotherapy*. (2nd ed.). Washington, DC: Hemisphere.

LUBORSKY, L., Crits-Christoph, P., McLellan, A. T., Woody, G., Piper, W., Liberman, B., Imber, S., & Pilkonis, P. (1986). The nonspecific hypothesis of therapeutic effectiveness: A current assessment. *American Journal of Orthopsychiatry, 56*, 501–512.

LYND, H. M. (1958). *On shame and the search for identity*. New York: Science Editions.

MAGER, R. F. (1992, April). No self-efficacy, no performance. *Training*, 32–36.

MAHALIK, J. R. (1990). Systematic eclectic models. *Counseling Psychologist, 18*, 655–679.

MAHONEY, M. J. (1991). *Human change processes*. New York: Basic Books.

MAHONEY, M. J., & Patterson, K. M. (1992). Changing theories of change: Recent developments in counseling. In S. D. Brown & R. W. Lent (Eds.), *Handbook of counseling psychology*. New York: Wiley.

MARKUS, H., & Nurius, P. (1986). Possible selves. *American Psychologist, 41*, 954–969.

MASLOW, A. H. (1968). *Toward a psychology of being*. (2nd ed.). New York: Van Nostrand Reinhold.

MATHEWS, B. (1988). The role of therapist self-disclosure in psychotherapy: A survey of therapists. *American Journal of Psychotherapy, 42(4)*, 521–531.

McFADDEN, L., Seidman, E., & Rappaport, J. (1992). A comparison of espoused theories of self- and mutual-help groups: Implications for mental health professionals. *Professional Psychology: Research and Practice, 23*, 515–520.

McMULLIN, R. E. (1986). *Handbook of cognitive therapy techniques*. New York: Norton.

McNEILL, B., & Stolenberg, C. D. (1989). Reconceptualizing social influence in counseling: The elaboration likelihood model. *Journal of Counseling Psychology, 36(1)*, 24–33.

McWHIRTER, E. H. (1991). Empowerment in counseling. *Journal of Counseling & Development, 69*, 222–227.

MEHRABIAN, A. (1971). *Silent messages*. Belmont, CA: Wadsworth.

MEHRABIAN, A., & Reed, H. (1969). Factors influencing judgments of psychopathology. *Psychological Reports, 24*, 323–330.

MILLER, G. A., Galanter, E., & Pribram, K. H. (1960). *Plans and the structure of behavior*. New York: Holt, Rinehart & Winston.

MILLER, L. M. (1984). *American spirit: Visions of a new corporate culture*. New York: Morrow.

MILLER, W. C. (1986). *The creative edge: Fostering innovation where you work*. Reading, MA: Addison-Wesley.

MULTON, K. D., Brown, S. D., & Lent, R. W. (1991). Relation of self-efficacy beliefs to academic outcomes: A meta-analytic investigation. *Journal of Counseling Psychology, 38(1)*, 30–38.

MURPHY, K. C., & Strong, S. R. (1972). Some effects of similarity self-disclosure. *Journal of Counseling Psychology, 19*, 121–124.

MYERS, J. E., Emmerling, D., & Leafgren, F. (Eds.). (1992). Wellness throughout the life span [Special issue]. *Journal of Counseling and Development, 71*.

NEIMEYER, G. (Ed.). (1993). *Constructivist assessment, a casebook*. Newbury Park, CA: Sage.

NEZU, A. M. (1985). Differences in psychological distress between effective and ineffective problem solvers. *Journal of Counseling Psychology, 32*, 135–138.

NORCROSS, J. C., Alford, B. A., & DeMichele, J. T. (1992). The future of psychotherapy: Delphi data and concluding observations. *Psychotherapy, 29*, 150–158.

NORCROSS, J. C., & Goldfried, M. R. (Eds.). (1992). *Handbook of psychotherapy integration*. New York: Basic Books.

NORCROSS, J. C., & Prochaska, J. O. (1988). A study of eclectic (and integrative) views revisited. *Professional Psychology: Research and Practice, 19*, 170–174.

NORCROSS, J. C., Strausser, D. J., & Faltus, F. J. (1988). The therapist's therapist. *American Journal of Psychotherapy, 42*, 53–66.

NORCROSS, J. C., & Wogan, M. (1987). Values in psychotherapy: A survey of practitioners' beliefs. *Professional Psychology: Research and Practice, 18(1)*, 5–7.

O'HANLON, W. H., & Weiner-Davis, M. (1989). *In search of solutions: A new direction in psychotherapy*. New York: Norton.

OKUN, B. F. (1990). *Seeking connections in psychotherapy*. San Francisco: Jossey-Bass.

O'LEARY, A. (1985). Self-efficacy and health. *Behaviour Research and Therapy, 23*, 437–451.

OTANI, A. (1989). Client resistance in counseling: Its theoretical rationale and taxonomic classification. *Journal of Counseling and Development, 67*, 458–461.

PANCOAST, D. L., Parker, P., & Froland, C. (Eds.). (1983). *Rediscovering self-help: Its role in social care: Vol 6. Social service delivery systems*. Beverly Hills, CA: Sage.

PATTERSON, C. H. (1985). *The therapeutic relationship: Foundations for an eclectic psychotherapy*. Pacific Grove, CA: Brooks/Cole.

PAYNE, E. C., Robbins, S. B., & Dougherty, L. (1991). Goal directedness and older-adult adjustment. *Journal of Counseling Psychology, 38(3)*, 302–308.

PEDERSEN, P. B. (1991a) Counseling international students. *The Counseling Psychologist, 19*, 10–58.

PEDERSEN, P. B. (1991b). Multiculturalism as a generic approach to counseling. *Journal of Counseling and Development, 70, (1)*, 6–12.

PEDERSEN, P. B. (Ed.) (1991c). Special issue: Multiculturalism as a fourth force in counseling. *Journal of Counseling and Development, 70, (1)*.

PERSONS, J. B. (1989). *Cognitive therapy in practice: A case formulation approach*. New York: Norton.

PETERSON, C., Seligman, M. E. P., & Vaillant, G. E. (1988). Pessimistic explanatory style as a risk factor for physical illness: A thirty-five-year longitudinal study. *Journal of Personality and Social Psychology, 55*, 23–27.

PHILLIPS, E. L. (1987). The ubiquitous decay curve: Service delivery similarities in psychotherapy, medicine, and addiction. *Professional Psychology: Research and Practice, 18*, 650–652.

POPE, K. S., & Vasquez, M. J. T. (1991). *Ethics in psychotherapy and counseling*. San Francisco: Jossey-Bass.

PROCTOR, E. K., & Rosen, A. (1983). Structure in therapy: A conceptual analysis. *Psychotherapy: Theory, Research, and Practice, 20*, 202–207.

PYSZCZYNSKI, T., & Greenberg, J. (1987). Self-regulatory perseveration and the depressive self-focusing style: A self-awareness theory of depression. *Psychological Bulletin, 102*, 122–138.

REANDEAU, S. G., & Wampold, B. E. (1991). Relationship of power and involvement to working alliance: A multiple-case sequential analysis of brief therapy. *Journal of Counseling Psychology, 38*, 107–114.

RIESSMAN, F. (1985). New dimensions in self-help. *Social Policy, 15(3)*, 2–4.

ROBERTSHAW, J. E., Mecca, S. J., & Rerick, M. N. (1978). *Problem-solving: A systems approach*. New York: Petrocelli Books.

ROBITSCHEK, C. G., & McCarthy, P. A. (1991). Prevalence of counselor self-reliance in the therapeutic dyad. *Journal of Counseling & Development, 69*, 218–221.

ROGERS, C. R. (1980). *A way of being*. Boston: Houghton Mifflin.

ROGERS, C. R., Perls, F., & Ellis, A. (1965). *Three approaches to psychotherapy 1 [Film]*. Orange, CA: Psychological Films.

ROGERS, C. R., Shostrom, E., & Lazarus, A. (1977). *Three approaches to psychotherapy 2 [Film]*. Orange, CA: Psychological Films.

ROSEN, S., & Tesser, A. (1970). On the reluctance to communicate undesirable information: The MUM effect. *Sociometry, 33*, 253–263.

ROSEN, S., & Tesser, A. (1971). Fear of negative evaluation and the reluctance to transmit bad news. *Proceedings of the 79th Annual Convention of the American Psychological Association, 6*, 301–302.

SARASON, I. G., Sarason, B. R., & Pierce, G. R. (1990). Social support: The search for theory. *Journal of Social and Clinical Psychology, 9*, 133–147.

SCHEIN, E. H. (1990, Spring). A general philosophy of helping: Process consultation. *Sloan Management Review*, 57–64.

SCHIFF, J. L. (1975). *Cathexis reader: Transactional analysis treatment of psychosis*. New York: Harper & Row.

SCHNEIDER, S. F. (1990). Psychology at the crossroads. *American Psychologist, 45*, 521–529.

SCHOEMAKER, P. J. H., & Russo, J. E. (1990). *Decision traps*. New York: Doubleday.

SCOGIN, F., Bynum, J., Stephens, G., & Calhoon, S. (1990). Efficacy of self-administered treatment programs: Meta-analytic review. *Professional Psychology: Research and Practice, 21(1)*, 42–47.

SELF-HELP REPORTER. (1985, Fall). Vol. 7(2). The self-help ethos.

SELF-HELP REPORTER. (1992, Summer), p. 1.

SELIGMAN, M. E. P. (1975). *Helplessness: On depression, development, and death*. San Francisco: W. H. Freeman.

SIEGMAN, A. W., & Feldstein, S. (Eds.). (1987). *Nonverbal behavior and communication* (2nd ed.). Hillsdale, NJ: Erlbaum.

SIMON, J. C. (1988). Criteria for therapist self-disclosure. *American Journal of Psychotherapy, 42(3)*, 404–415.

SMABY, M., & Tamminen, A. W. (1979). Can we help belligerent counselees? *Personnel and Guidance Journal, 57*, 506–512.

SMITH, M. L., Glass, G. V., & Miller, T. I. (1980). *The benefits of psychotherapy*. Baltimore: Johns Hopkins University Press.

SNYDER, C. R. (1984, September). Excuses, excuses. *Psychology Today*, pp. 50–55.

SNYDER, C. R., & Higgins, R. L. (1988). Excuses: Their effective role in the negotiation of reality. *Psychological Bulletin, 104,* 23–35.

SNYDER, C. R., Higgins, R. L., & Stucky, R. J. (1983). *Excuses: Masquerades in search of grace.* New York: Wiley.

SPEIGHT, S. L., Meyers, L. J., Cox, C. I., & Highlen, P. S. (1991). A redefinition of multicultural counseling. *Journal of Counseling & Development, 70,* 29–36.

STEIN, B. A. (1980). *Quality of work life in context: What every practitioner should know.* Unpublished manuscript.

STEKETEE, G., & Chambless D. L. (1992). Methodological issues in prediction of treatment outcome. *Clinical Psychological Review, 12(4),* 387–400.

STRICKER, G., & Fisher, A. (Eds.). (1990). *Self-disclosure in the therapeutic relationship.* New York: Plenum.

STRONG, S. R. (1968). Counseling: An interpersonal influence process. *Journal of Counseling Psychology, 15,* 215–224.

STRONG, S. R., & Claiborn, C. D. (1982). *Change through interaction: Social psychological processes of counseling and psychotherapy.* New York: Wiley.

SUE, D. W. (1990). Culture-specific strategies in counseling: A conceptual framework. *Professional Psychology: Research and Practice, 21(6),* 424–433.

SUE, D. W., Arredondo, P., & McDavis, R. J. (1992). Multicultural counseling competencies and standards: A call to the profession. *Journal of Counseling & Development, 70,* 477–486.

SYKES, C. J. (1992). *A nation of victims.* New York: St. Martin's.

TAUSSIG, I. M. (1987). Comparative responses of Mexican Americans and Anglo-Americans to early goal setting in a public mental health clinic. *Journal of Counseling Psychology, 34,* 214–217.

TESSER, A., & Rosen, S. (1972). Similarity of objective fate as a determinant of the reluctance to transmit unpleasant information: The MUM effect. *Journal of Personality and Social Psychology, 23,* 46–53.

TESSER, A., Rosen, S., & Batchelor, T. (1972). On the reluctance to communicate bad news (the MUM effect): A role play extension. *Journal of Personality, 40,* 88–103.

TESSER, A., Rosen, S., & Tesser, M. (1971). On the reluctance to communicate undesirable messages (the MUM effect): A field study. *Psychological Reports, 29,* 651–654.

TRACEY, T. J. (1991). The structure of control and influence in counseling and psychotherapy: A comparison of several definitions and measures. *Journal of Counseling Psychology, 38,* 265–278.

TYLER, F. B., Pargament, K. I., & Gatz, M. (1983). The resource collaborator role: A model for interactions involving psychologists. *American Psychologist, 38,* 388–398.

TYRON, G. S., & Kane, A. S. (1993). Relationship of working alliance to mutual and unilateral termination. *Journal of Counseling Psychology, 40(1),* 33–36.

VACHON, D.O., & Agresti, A. A. (1992). A training proposal to help mental health professionals clarify and manage implicit values in the counseling process. *Professional Psychology: Research and Practice, 23,* 509–514.

WAHLSTEN, D. (1991). Nonverbal behavior and self-presentation. *Psychological Bulletin, 110,* 587–595.

WATCHEL, P. L. (1989, August 6). Isn't insight everything? [Book Review]. *New York Times,* p. 18.

WATKINS, C. E., Jr. (1990). The effects of counselor self-disclosure: A research review. *The Counseling Psychologist, 18(3),* 477–500.

WATKINS, C. E., Jr., & Schneider, L. J. (1989). Self-involving versus self-disclosing counselor statements during an initial interview. *Journal of Counseling and Development*, *67*, 345–349.

WATSON, D. L., & Tharp, R. G. (1993). *Self-directed behavior* (6th ed.). Pacific Grove, CA: Brooks/Cole.

WEICK, K. E. (1979). *The social psychology of organizing* (2nd ed.). Reading, MA: Addison-Wesley.

WEINRACH, S. G. (1989). Guidelines for clients of private practitioners: Committing the structure to print. *Journal of Counseling and Development*, *67*, 299–300.

WEISZ, J. R., Rothbaum, F. M., & Blackburn, T. C. (1984). Standing out and standing in: The psychology of control in America and Japan. *American Psychologist*, *39*, 955–969.

WEISZ, J. R., Weiss, B., & Donenberg, G. R. (1992). The lab versus the clinic. *American Psychologist*, *47(12)*, 1578–1585.

WESTERMAN, M. A. (Ed.). (1989). Putting insight to work [Special section]. *Journal of Integrative and Eclectic Psychotherapy*, *8*, 195–250.

WHEELER, D. D., & Janis, I. L. (1980). *A practical guide for making decisions*. New York: Free Press.

WIENER, M., Budney, S., Wood, L., & Russel, R. L. (1989). Nonverbal events in psychotherapy. *Clinical Psychology Review*, *9*, 487–504.

WILLIAMS, M. H. (1985). The bait-and-switch in psychotherapy. *Psychotherapy*, *22(1)*, 110–113.

WILLIAMS, R. (1989, January-February). The trusting heart. *Psychology Today*, pp. 36–42.

WINSLOW, R. (1993, January 7). Integra Inc. develops system to assess effectiveness of psychiatric treatment. *The Wall Street Journal*, p. B3.

WOODY, R. H. (1991). *Quality care in mental health*. San Francisco: Jossey-Bass.

YANKELOVICH, D. (1992, October 5). How public opinion really works. *Fortune*, pp. 102–108.

ZANE, N. W. S., Sue, S., Hu, L., & Kwon, J. (1991). Asian-American assertion: A social learning analysis of cultural differences. *Journal of Counseling Psychology*, *38(1)*, 63–70.

Name Index

Subject Index

Abortion, 179
Acceptability
  of costs for residential program, 294–295
  of strategy benefits, 293–294
Accommodating mode, 75–76
Accomplishment
  goal as, 256–257
  of goals
    brainstorming and, 280–282
    realistic time frames for, 264–265
Accurate empathy, 115–116
Action
  as challenge goal, 171–172
  of client, 29
  delaying, 279–280
  empathic, 107
  faces of, 71–78
  goal-accomplishing, 24
    helping clients engage in, 314–315
  goals and, 221
  goal setting and, 221–224
  imprudent, avoidance of, 278–279
  influencing mode vs. accommodating
    mode, 75–76
  informal vs. formal, 74–75
  inside and outside helping sessions, 73–74
  internal vs. external, 71–72, 201
  linking
    to challenge, 172–173
    to new perspective, 160–161
    to Step III-B, 298–299
    to Step III-C, 309
    to strategy formulation, 284–286
  plan of, 302–303
  in Stage II, 273
  in Step I-A, 153–155
  in Step I-C, 214–215
  stimuli for, goals as, 221–222
  storytelling and, 137
  sustained
    incentives and rewards for, 319–320
    skills for, 320–321
  understanding and, 41
  unproductive vs. productive, 76–78
Action-based contracts, developing, 321–322
Action strategies
  brainstorming, 34–35
  development of, 23–24

Action Strategies *(continued)*
  identifying, probes and prompts for,
    282–283
  implementation of, 38–39
Active listening, 94–100
Activities, vs. outcome and impact, 234–235
Adaptive learning, 158
Addicts, working with, 180
Advanced empathy, 180–184, 197
Advice, as helper response, 119
Affect, 68, 70–71
Agency, sense of, 82–84
Agendas
  commitment to, 266–272
  competing, dealing with, 271–272
  productive, crafting, 255
  in program focus, 54–55
Aggression, 164
Agreement, sympathy and, 120
AIDS, 179, 203, 260
Alcohol addiction counseling, helper self-
    disclosure and, 184
Ambiguity
  acceptance of, 226
  encouragement of clarity and, 140
Analysis
  force-field, 317–319, 327
  in rational decision making, 201
Anger, hurt and, 180
Anxiety reduction, 83
Apologetic challenge, 193
Appeal, of goals, 268–269
Assertiveness
  in communication, 109
  of helpers, 51
  results of, 6–7
  self-efficacy and, 82
Assessment
  initial, 144–145
  of learning abilities, 145
  ongoing, evaluation questions for, 156
  as search for client resources, 145–147
  in step I-A, 143–147
Attending
  definition of, 90
  function of, 91
  microskills of, 91–93
  nonverbal communication and, 93

TO THE OWNER OF THIS BOOK:

We hope that you have found *The Skilled Helper,* Fifth Edition, useful. So that this book can be improved in a future edition, would you take the time to complete this sheet and return it? Thank you.

School and address: _____

Department: _____

Instructor's name: _____

1. What I like most about this book is: _____

_____

2. What I like least about this book is: _____

_____

3. My general reaction to this book is: _____

_____

4. The name of the course in which I used this book is: _____

_____

5. Were all of the chapters of the book assigned for you to read? _____

   If not, which ones weren't? _____

_____

6. In the space below, or on a separate sheet of paper, please write specific suggestions for improving this book and anything else you'd care to share about your experience in using the book.

_____

_____

_____

_____

_____

_____

Optional:

Your name: _____ Date: _____

May Brooks/Cole quote you either in promotion for *The Skilled Helper*, Fifth Edition, or in future publishing ventures?

Yes: _____ No: _____

Sincerely,

Gerard Egan

Brooks/Cole is dedicated to publishing quality publications for education in the human services fields. If you are interested in learning more about our publications, please fill in your name and address and request our latest catalogue.

Name: _____

Street Address: _____

City, State, and Zip: _____

FOLD HERE

BUSINESS REPLY MAIL

FIRST CLASS   PERMIT NO. 358   PACIFIC GROVE, CA

POSTAGE WILL BE PAID BY ADDRESSEE

ATT: *Human Services Catalogue*

Brooks/Cole Publishing Company
511 Forest Lodge Road
Pacific Grove, California 93950-9968

FOLD HERE